83 Advances in Polymer Science

Biopolymers

With Contributions by
A. H. Clark, K. Kamide,
S. B. Ross-Murphy, M. Saito

With 114 Figures and 15 Tables

Springer-Verlag Berlin Heidelberg GmbH

ISBN 978-3-662-15168-6 ISBN 978-3-540-47827-0 (eBook)
DOI 10.1007/978-3-540-47827-0

Library of Congress Catalog Card Number 61-642

Originally published by Springer-Verlag Berlin Heidelberg New York in 1987.
Softcover reprint of the hardcover 1st edition 1987

Typesetting and Offsetprinting: Th. Müntzer, GDR;

2154/3020-543210

This volume is affectionately dedicated to Professor Manfred Gordon on the occasion of his 70th birthday for his great contribution to science and humanity

Reproduced with kind permission of the Artist.

Gerda Stevenson

Editors

Table of Contents

Cellulose and Cellulose Derivatives: Recent Advances in Physical Chemistry

Kenji Kamide and Masatoshi Saito
Fundamental Research Laboratory of Fibers and Fiberforming Polymers, Asahi Chemical Industry, Co. Ltd., Takatsuki, Osaka 569, Japan

This article reviews a recent progress on the physical chemistry of cellulose and cellulose derivatives (CD) and their applications to some industrial fields. Average degree of substitution for each hydroxyl groups attached to carbon 2, 3, and 6 in a pyranose ring $\langle\langle f_k \rangle\rangle$ *(*$k = 2, 3, 6$*) could be estimated by* ^1H*- and* ^{13}C*-NMR methods and distribution of total degree of substitution of some CD was evaluated by thin-layer chromatography.* $\langle\langle f_k \rangle\rangle$ *correlated closely with the anticoagulant activity of sodium cellulose sulfate and also with the absorbency against aqueous liquid of sodium carboxymethyl cellulose. Successive solution fractionation method afforded us to prepare CD samples with relatively narrow molecular weight distribution. Light scattering measurements on the gel-free CD solutions were carried out and the number-average molecular weight of cellulose acetate (CA) was determined by membrane and vapor pressure osmometry and gel-permiation chromatography. Lower and upper critical solution temperatures were determined for CA-solvent systems. The pore forming mechanism in the CA-solvent casting process was discribed relating to critical phenomena. The solvation was verified by the chemical shift in NMR spectra and by the adiabatic compressibility. The significant contribution of the draining effect on the hydrodynamic properties was experimentally confirmed. The excluded volume effect in CA solutions was very small. The rigidity of CD molecules in the unperturbed state was estimated by various methods based on the pearl necklace and wormlike chain models. The unperturbed chain dimension of CA molecules in the solutions was decided by the polarity of the solvent and the total degree of substitution. Cellulose dissolved in a hypothetical non-polar solvent behaves as almost a freely rotating chain and the low degree of flexibility of the cellulose chain deduced from the physical properties of cellulose solution and solid, is caused by the solvation or intra-or inter-molecular hydrogen bond. The solubility of cellulose in the aqueous alkali solution was discussed.*

Advances in Polymer Science 83

List of Symbols and Abbreviations

A	unperturbed chain dimension
A′	absorbency
$A_{c,w}$	weight-average combined acetic acid content
A_f	unperturbed chain dimension of a hypothetical chain with free internal rotation
A_2	second virial coefficient
A_∞	unperturbed chain dimension of a chain with infinite molecular weight
B	long range interaction parameter
C_∞	characteristic ratio
D_0	diffusion constant
⟪F⟫	average total degree of substitution per glucopyranose ring
L	contour length
LD_{50}	acute toxicity
M_b	mean molecular weight per skeletal bond
M_n	number-average molecular weight
M_v	viscosity-average molecular weight
M_w	weight-average molecular weight
M_0	a parameter depending on the molecular weight range in which Mark-Houwink-Sakurada equation is valid
N	total number of molecules in a sample
Nc	nitrogen content of a sample
N′	number of segments in a molecular chain
N_A	Avogadro number

P	a parameter related to the frictional coefficient ξ, and analogous to Flory's viscosity parameter Φ
$\langle R^2\rangle^{1/2}$	mean square end-to-end distance
R_f	rate of flow
Sa	solubility of cellulose in aqueous alkali solution
$\langle S^2\rangle^{1/2}$	mean square radius of gyration
T	temperature
T_c	critical solution temperature
ΔT_s	temperature difference between solution and solvent in the cell of vapor pressure osmometer
X	draining parameter
X_w	weight-average molar volume ratio of polymer to solvent
X_z	z-average molar volume ratio of polymer to solvent
a'	length of a link in pearl neck-lace model
c	polymer concentration, g/cm^3
d	hydrodynamic radius of a segment
$\langle\langle f_k\rangle\rangle$	average degree of substitution per hydroxyl group in a pyranose ring
$m_{p(s)}$	molecular weight of the repeating unit of a polymer (solvent)
m_0	molecular weight of a segment
n	number of grams of the solvating solvent molecules per 1 g of polymer
p_j	degree of polymerization of j-th polymer
p_1	concentration dependence coefficient of polymer volume fraction v_p, in polymer-solvent interaction parameter χ
p_2	concentration dependence coefficient of v_p^2 in χ
q_{BD}	persistence length determined by Benoit-Doty method
q_{YF}	persistence length determined by Yamakawa-Fujii method
q_{BD}^0	q_{BD} in unperturbed state
q_{CL}^0	unperturbed persistence length at coil limit
s_0	sedimentation constant
s_0'	number of the solvating solvent molecules per repeating unit at infinite dilution
v_p	polymer volume fraction, v/v
$\bar{v}_p$	molar volume of polymer
v_p^c	critical polymer concentration, v/v
v_p^0	initial polymer concentration before phase separation
w_2	weight fraction of polymer, w/w
a	exponent in Mark-Houwink-Sakurada equation
a_d	exponent in the equation representing the dependence of diffusion coefficient on molecular weight
a_s	exponent in the equation representing the dependence of sedimentation coefficient on molecular weight
a_Φ	exponent in the equation representing the dependence of Flory's viscosity parameter Φ on molecular weight
a_ξ	exponent in the equation representing the dependence of the frictional coefficient on molecular weight
a_1	exponent in the equation representing the dependence of the linear expansion coefficient a_s on molecular weight

a_2	exponent in the equation representing the dependence of the ratio square of radius of gyration in unperturbed state to molecular weight $\langle S^2\rangle_0/M$ on molecular weight
Φ	Flory's viscosity parameter
ψ	penetration function
α_s	linear expansion coefficient
β	adiabatic compressibility
β'	binary cluster integral
γ	correlation coefficient
δ	chemical shift
δ^-	electronegativity
ε	dielectric constant
$[\eta]$	limiting viscosity number (intrinsic viscosity)
η_0	viscosity of solvent
θ	Flory's theta solvent
λ	exponent in the equation representing the dependence of radius of gyration on molecular weight
ξ	frictional coefficient
ϱ	density of the solution
σ	conformation parameter
χ	polymer-solvent interaction parameter
χ'	anti-coagulant activity
χ_0	concentration independent part of χ
χ_0^c	critical χ_0
$\chi_{am}(X)$	amorphous content determined as $1 - \chi_c$ (χ_c = cristallinity determined by X-ray diffraction method)
$\chi_{ac}(IR)$	fraction of accessible part at equilibium determined by the deuteration IR method
$\chi_h(NMR)$	relative amount of higher field peaks of the C_4 carbon peaks in NMR spectrum

1 Introduction

Cellulose is a naturally occurring and reproducible polymer of β-D-glucopyranose (Fig. 1). Cellulose and its derivatives, in which the hydroxyl groups of cellulose are partly or fully substituted with other functional groups, have been found wide industrial applications, such as in paper, fibers, plastic, membranes, food, and so on. The study of the number-average molecular weight M_n and the limiting viscosity number $[\eta]$ of cellulose derivatives (CD) in solution started as early as in the 1930's [1], and this kind of work contributed very much to the establishment of the concept of "macromolecular compound" by Staudinger. Nevertheless, very few comprehensive and reliable studies on the molecular characteristics of CD have been reported, because of the experimental difficulties met in the determination of the distribution of the substituent groups, fractionation, and removal of gel-like materials from the solution. Moreover, the experimental data, unsystematically obtained for some CD in solution, could not be explained reasonably in terms of the known solution theories, applicable to many synthetic flexible polymers [2]. For this reason, the molecular characterization of cellulose and CD remained unsolved even in the 1970's.

In this article, we review the recent progress on the molecular characterization of cellulose and CD.

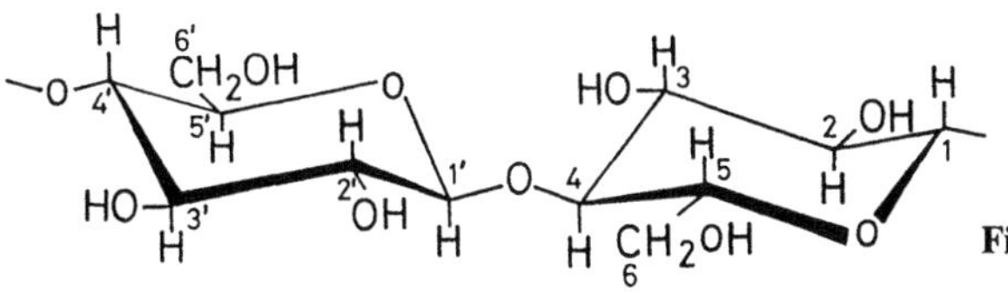

Fig. 1. Steric configuration of cellulose

2 Distribution of Degree of Substitution in Cellulose Derivatives

As is well known, the glucopyranose unit constituting a cellulose molecule has one primary and two secondary hydroxyl groups at the C_2, C_3, and C_6 positions. The extent of substitution of these hydroxyl groups with substituent groups may differ with different C positions, due to different reactivity. The average degree of substitution for each hydroxyl group ($\langle\langle f_k \rangle\rangle$, k = 2, 3, 6 Eq. (4)) is an important parameter for characterizing cellulose derivatives. With the help of three parameters, the molecular properties of cellulose derivative (CD) polymers can be well understood.

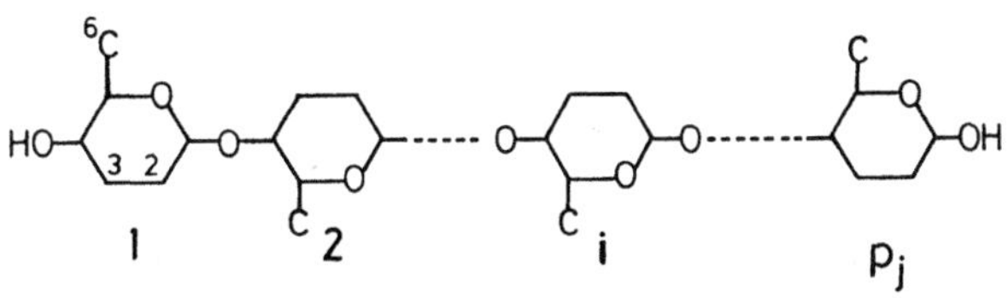

Fig. 2. Schematic representation of the j-th cellulose molecule having the degree of polymerization p_j [3]

Kamide and Okajima [3] defined the probability of substitution of the hydroxyl group, attached to the C_k (k = 2, 3, 6) atoms of the i-th glucopyranose unit in the j-th molecule, with the functional group by $f^j_{2,i}$, $f^j_{3,i}$, $f^j_{6,i}$, respectively (Fig. 2), and designated the total degree of substitution (DS) of the i-th ring in the j-th molecule by F^j_i. The following relations are readily derived [3].

$$f^j_{2,i} + f^j_{3,i} + f^j_{6,i} = F^j_i \quad (1)$$

$$\sum_{i=1}^{p_j} F^j_i / p_j = \langle F^j \rangle \quad (2)$$

$$\sum_{j=1}^{N} p_j \langle F^j \rangle \Big/ \sum_{j=1}^{N} p_j = \langle\langle F \rangle\rangle \quad (3)$$

$$\sum_{j=1}^{N} \sum_{i=1}^{p_j} f^j_{k,i} \Big/ \sum_{j=1}^{N} p_j = \langle\langle f_k \rangle\rangle \qquad k = 2, 3, 6 \quad (4)$$

$$\langle\langle f_2 \rangle\rangle + \langle\langle f_3 \rangle\rangle + \langle\langle f_6 \rangle\rangle = \langle\langle F \rangle\rangle \quad (5)$$

where N is the total number of molecules, p_j the degree of polymerization of the j-th polymer molecule.

(a) Chemical Methods and NMR

Several attempts have been made to evaluate $\langle\langle f_k \rangle\rangle$ in a trihydric alcohol unit of cellulose acetate (CA) by chemical [4,5] and NMR methods [3,6]. The former are generally based on the difference in reactivity of the three unsubstituted hydroxyl groups in tosylation [4] with p-toluenesulfonyl chloride, followed by iodination with sodium iodide, or on tritylation [5] with trityl chloride in the presence of pyridine. These chemical methods are tedious and time-consuming. Moreover, the degree of substitution at C_2 and C_3 cannot be evaluated separately. These methods are relative, requiring the total DS value which has to be determined by another method. In contrast to this, the evaluation by NMR is an absolute method, however unfortunately it is applicable only to completely or almost completely substituted CA. Otherwise, the peaks due to the O-acetyl protons cannot be resolved into the components corresponding to the C_k positions, since a large number of glucopyranose units with magnetically unequivalent O-acetyl groups are possible. To avoid this, Goodlett et al. [6] have acetylated incompletely substituted CA polymer with deuterated acetyl chloride, before ^{1}H-NMR measurements.

Recently, Kamide et al. [3,7] determined the $\langle\langle f_k \rangle\rangle$ of CA with $\langle\langle F \rangle\rangle$ ranging from 0.49 to 2.92 (Table 1). The $\langle\langle f_k \rangle\rangle$ values of CA(2.92) (the number in parentheses denotes $\langle\langle F \rangle\rangle$) determined by ^{1}H- and ^{13}C-NMR, agree fairly well between themselves and with the chemical analysis ($\langle\langle F \rangle\rangle$).

Wu [8] analyzed the distribution of substituted glucopyranose units in cellulose nitrate (CN) by ^{1}H- and ^{13}C-NMR methods. The $\langle\langle f_k \rangle\rangle$ values of sodium cellulose sulfate (NaCS) [9], and sodium carboxymethylcellulose (NaCMC) [10] were determined using ^{13}C-NMR by Kamide et al.

Table 1. Evaluation of ⟪f_k⟫ for cellulose acetate (CA) by chemical analysis, ^{1}H- and ^{13}C-NMR measurements [7]

Polymer	Solvent	⟪F⟫ chemical analysis	position	^{1}H-NMR		^{13}C-NMR	
				O-acetyl proton peak (ppm)	⟪f_k⟫[a]	Carbonyl carbon peak (ppm)	⟪f_k⟫[b]
CA(2.92)	Trichloro-methane-d_1	2.92	C_6	2.13	0.99	170.2	1.00
			C_2	2.02	1.01	169.7	1.02
			C_3	1.98	1.092	169.3	0.89
CA(2.46)	Acetone-d_6	2.46	C_6	2.07	—	170.7	0.82
			C_2	1.97	—	170.1	0.75
			C_3	—	—	169.7	0.89
CA(1.75)	Dimethyl-sulphoxide-d_6	1.75	C_6	—	—	170.5	0.59
			C_2	—	—	169.6	0.53
			C_3	—	—	169.3	0.63
CA(0.49)	Dimethyl-sulphoxide-d_6	0.49	C_6	—	—	169.9	0.19
			C_2	2.03	—	169.1	0.10
			C_3	—	—	168.8	0.20

[a] Estimated from the peak area ratios calculated from a triangle approximation; [b] estimated from integrated peak intensity ratios

(b) Thin-layer Chromatography (TLC)

This method gives information on the distribution of ⟨F^j⟩, (g(⟨F^j⟩)), together with ⟪F⟫. Kamide et al. estimated g(⟨F^j⟩) and ⟪F⟫ on numerous CA [11,12] and CN [13] samples. For CA, the following relations between the weight-average rate of flow $R_{f,w}$, the weight-average M_w, and the number average M_n of the molecular weight have been established empirically [12].

$$R_{f,w} = 3.51 - 5.4\times10^{-2}A_{c,w} - 3.66\times10^{-8}M_w + 1.13\times10^3 M_n^{-1} \quad (6)$$

($A_{c,w}$ = weight average of the total acetic acid content).

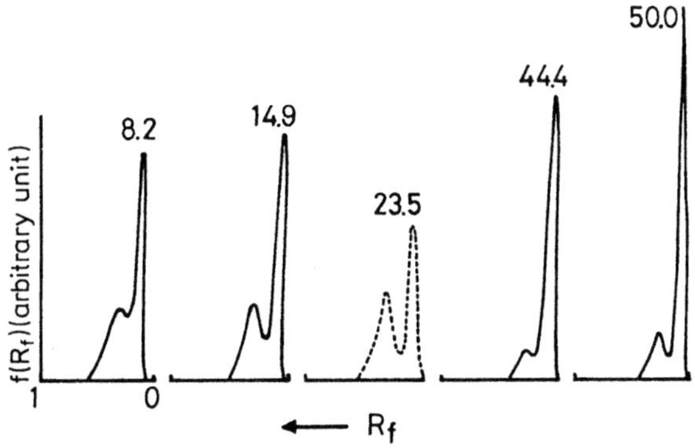

Fig. 3. Typical TLC chromatograms of cellulose triacetate (CTA) fractions and whole polymer ($A_{c,w}$ = 61.0 wt %) having various M_w [12]; solid lines: fractions; broken line: whole polymer; numbers on curves represent 10^{-4} M_w. (R_f = rate of flow.)

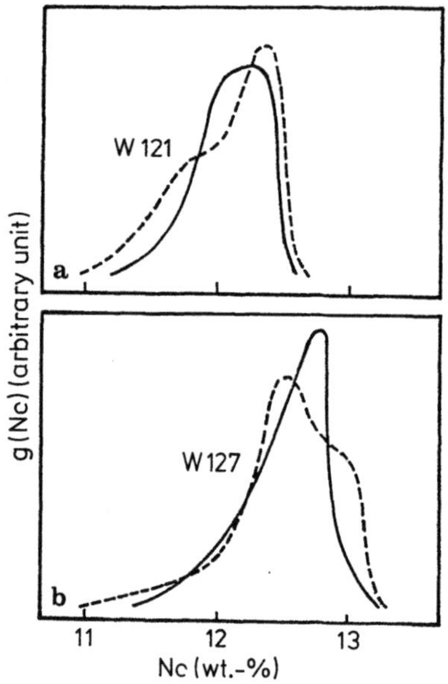

Fig. 4a and b. Differential weight distribution of nitrogen content (Nc by wt %) g(Nc) of cellulose nitrate whole polymer with average Nc = 12.1 wt% (W121) and whole polymer with average Nc = 12.7 wt% (W127), as estimated by TLC [13]. Solid line: silica gel, open TLC, nitromethanemethanol (20/80, v/v); broken line: kieselguhr, VP-TLC, acetone-methanol-chloroform (10/10/5, v/v/v)

The chromatograms of cellulose triacetate (CTA) whole polymer ($A_{c,w}$ = 61.0 wt %, dotted curve) and its fractions (solid curves) are illustrated in Fig. 3. For the cellulose diacetate (CDA) and CTA fractions, the TLC becomes apparently sharp with an increase in M_w. The double-peaked form of the chromatograms is characteristic of the CTA samples, although their gel permeation chromatography (GPC) curves have been found to be single-peaked. This fact implies that the peak at the lower end of R_f corresponds to fully substituted CTA and the peak at the higher end is obviously due to the existence of not-fully substituted acetate. In this sense, real CTA is a binary mixture of ideal CTA and CDA.

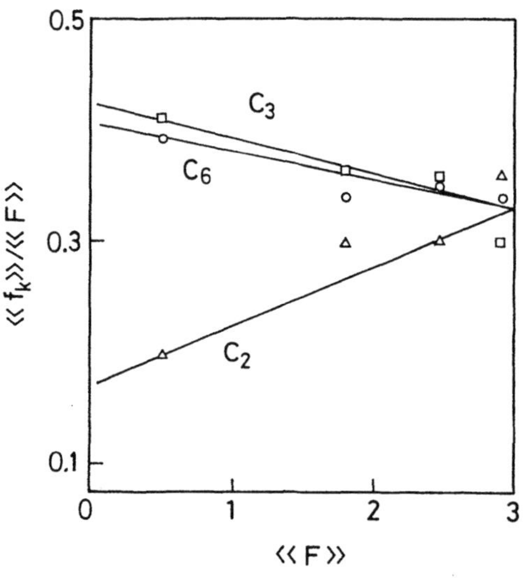

Fig. 5. Plot of the ratio of the average probability of substitution of the OH group attached to the C_k (k = 2, 3, 6) position, $\langle\langle f_k \rangle\rangle$, to the total degree of substitution of OH groups of the cellulose acetate molecule $\langle\langle F \rangle\rangle$ ($\langle\langle f_k \rangle\rangle / \langle\langle F \rangle\rangle$) as a function of $\langle\langle F \rangle\rangle$ [7]. ○: k = 6; △: k = 2; □: k = 3 ($\langle\langle f_k \rangle\rangle$ determined by ^{13}C NMR.)

Figure 4 shows the differential weight distribution of the nitrogen content Nc, g(Nc), for two CN samples, as estimated by two TLC techniques [13]. The full lines are obtained by open TLC with nitromethane-methanol (20/80, v/v), and the broken lines by VP-TLC with acetone-methanol-chloroform (10/10/5, v/v/v). The agreement between the g(Nc) curves evaluated by both methods is fair, considering the possibility of large experimental uncertainty.

(c) Correlation Between Distribution of Degree of Substitution and Other Physical Properties

i) Change in $\langle\langle f_k \rangle\rangle$ *During Hydrolysis of CA*

Figure 5 shows the variation of the ratio $\langle\langle f_k \rangle\rangle / \langle\langle F \rangle\rangle$ (k = 2, 3, 6) during the hydrolysis of CA [7]. As the hydrolysis proceeds, the ratios $\langle\langle f_6 \rangle\rangle / \langle\langle F \rangle\rangle$ and $\langle\langle f_3 \rangle\rangle / \langle\langle F \rangle\rangle$ increase, but $\langle\langle f_2 \rangle\rangle / \langle\langle F \rangle\rangle$ decreases significantly, indicating preferential removal of the acetyl group from the C_2 position.

ii) Anticoagulant Activity of NaCS

The chemical structure of cellulose sulfate resembles that of muco-polysaccharides, such as heparin and condroitin sulfate, which are now in wide use as naturally occurring blood anti-coagulants. An anti-coagulant activity of NaCS was first reported by Bergström [14] as early as 1935, and thereafter the Biological Institute of the Carlsberg Foundation in Copenhagen demonstrated that the coagulation time of whole blood increased by the addition of cellulose trisulfate and other polysaccharide sulfates such as amylose sulfate and amylopectin sulfate [15–19]. Felling and Wiley [20], Rothschild [21], and Rothschild and Castania [22] found many interesting pharmaceutical characteristics of NaCS including inhibitory action to pancreatic ribonuclease [20], kininogen depleting action [21], and endotoxin shock in dogs by the treatment with NaCS [22]. The aminosulfate group at the C_2 position of uronic units in heparin was found to play an important role in anti-coagulant activity [23]. Desulfation of this aminosulfate group was found to lower the anti-coagulant activity of heparin. These experimental results on heparin strongly suggest that the physiological activity, including anti-coagulant activity, of polymers is closely related to their molecular characteristics. Kamide et al. [24] disclosed for NaCS that its anti-coagulant and other pharmaceutical activities are influenced by molecular parameters such as molecular weight, chain structure, distribution of substituent groups along the molecular chain, and probability of substitution at the C_2, C_3, and C_6 positions of the glucopyranose units ($\langle\langle f_k \rangle\rangle$, k = 2, 3, 6) and $\langle\langle F \rangle\rangle$.

Figures 6 and 7 show the anti-coagulant activity χ', expressed in international units (IU)/mg, as determined by the modified Commentary of the Japanese Pharmacopoeia [24], and the acute toxicity LD_{50} (when anti-coagulant is injected into the vein of a rat), as functions of ($\langle\langle f_2 \rangle\rangle + \langle\langle f_3 \rangle\rangle$) and M_n [24]. The solid lines represent the contours of the same χ' (Fig. 6) and LD_{50} (Fig. 7). Larger ($\langle\langle f_2 \rangle\rangle + \langle\langle f_3 \rangle\rangle$) results in larger χ' when compared at the same M_n. For a fixed ($\langle\langle f_2 \rangle\rangle + \langle\langle f_3 \rangle\rangle$) χ' tends to increase with a decrease in M_n. From the anti-coagulant activity point of view, it is highly desirable to prepare NaCS with a higher ($\langle\langle f_2 \rangle\rangle + \langle\langle f_3 \rangle\rangle$) value and lower M_n. When compared at the same M_n, LD_{50} decreases as ($\langle\langle f_2 \rangle\rangle + \langle\langle f_3 \rangle\rangle$) increases. This condition causes

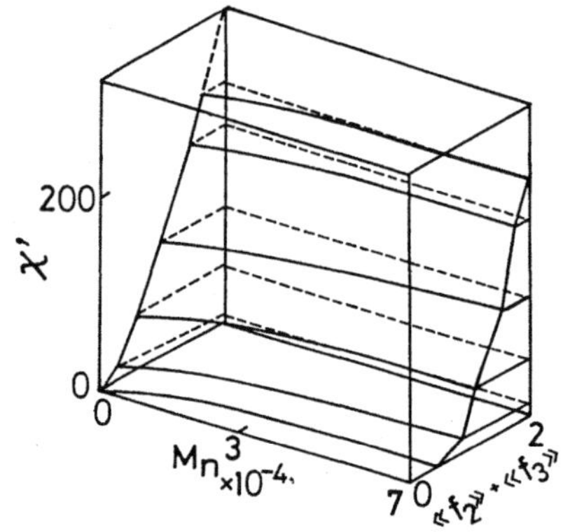

Fig. 6.

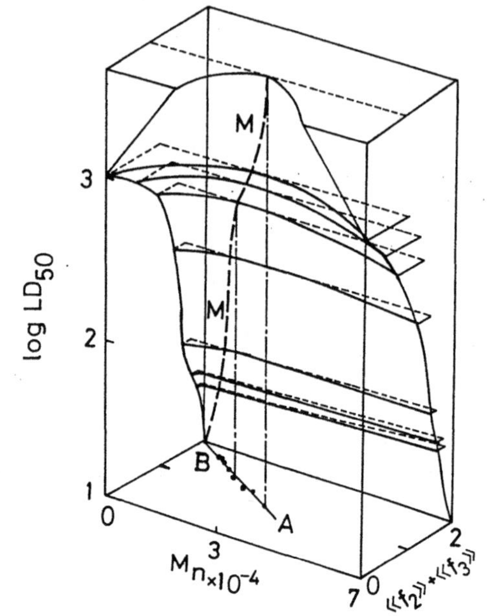

Fig. 7.

Fig. 6. Anticoagulant activity χ' plotted versus number average molecular weight M_n and the sum of the degree of substitution at carbon positions 2 and 3, $\langle\langle f_2 \rangle\rangle + \langle\langle f_3 \rangle\rangle$ [24)]

Fig. 7. Plots of acute toxicity LD_{50} of NaCS versus M_n and $\langle\langle f_2 \rangle\rangle + \langle\langle f_3 \rangle\rangle$ [24)]. The line M represents maximum values of $\langle\langle f_2 \rangle\rangle + \langle\langle f_3 \rangle\rangle$ for a given LD_{50}. AB denotes the projection of the line M on the plane $LD_{50} = 0$

trouble in obtaining NaCS with a high anti-coagulant activity and a desirable LD_{50}. The line M represents maximum acceptable ($\langle\langle f_2 \rangle\rangle + \langle\langle f_3 \rangle\rangle$ for a NaCS sample, hence the maximum anti-coagulant activity for a given LD_{50}. The next to best molecular and structual parameters of NaCS as anti-coagulants can be estimated: Fixing LD_{50} at the level of heparin, NaCS with $M_n = 1.65 \times 10^4$ and $(\langle\langle f_2 \rangle\rangle + \langle\langle f_3 \rangle\rangle) = 1.23$ (so that χ' is 120 IU/mg) is acceptable. If χ' is desired to be 152/mg, $M_n = 2.4 \times 10^4$ and $(\langle\langle f_2 \rangle\rangle + \langle\langle f_3 \rangle\rangle) = 1.4$ are relevant and result in $LD_{50} = 320 - 460$. It should be noted here that the oral LD_{50} of NaCS for a rat was found to be larger than 1.5×10^4 mg $\cdot$ kg^{-1}, and hence presents no toxic problem.

The effect of NaCS on all coagulation factors (I—XII) is at least the same as that of heparin. In fact NaCS inhibits the action of the anticoagulant factor VIII (anti-thermophilic globulin) much more effectively than heparin. Therefore, it may be concluded that NaCS acts mainly on factor VIII and partly on factor IX (plasma thromboplastic component); the latter was found to be suppressed by heparin.

Certain pharmacodynamic properties of NaCS were evaluated using rats, rabbits, dogs, and cats. Administration of 1 mg $\cdot$ kg^{-1} of two NaCS samples to dogs had no influence on blood pressure, heart rate, or respiration [25)]. Because 1×10^4 IU/mg of heparin is usually used for blood-dialysis for chronical renal insufficiency, about 1 mg $\cdot$ kg^{-1}, which was found to have no effect on the cardiac function of dogs, may have no influence on that on human beings.

iii) Effect of $\langle\langle f_k \rangle\rangle$ *of NaCMC on its Absorbency Towards Aqueous Liquids*

Jansen [26] was the first who synthesized sodium carboxymethylcellulose (NaCMC) in 1918 and invented a manufacturing process using the reaction of alkalicellulose with sodium monochloroacetate. Since then, NaCMC with $\langle\langle F \rangle\rangle$ of 0.5–1.0 was commercialized worldwide in large scale to find numerous industrial applications. However the characteristic features, including the solubility in water and aqueous salt solutions, and the interaction with cationic compounds, have not been thoroughly discussed, and only in a non-systematic manner [27,28]. Recently, Kamide et al. [10] reported a paricularly high degree of absorbing power of NaCMC toward various aqueous liquids, and interpreted the absorbency in terms of $\langle\langle f_k \rangle\rangle$ and $\langle\langle F \rangle\rangle$.

Figure 8 shows the relations between the absorbency A′ towards pure water (a), aq. NaCl (b), aq. $CaCl_2$ (c), aq. $AlCl_3$ (d), and $\langle\langle F \rangle\rangle$ for NaCMC from cellulose I and cellulose II (cf. Fig. 10) [10]. NaCMC synthesized from cellulose II has a great advantage as an absorbing agent, especially for aq. NaCl or anything similar. As the average molecular weight increases, A′ for a given $\langle\langle F \rangle\rangle$, as well as the maximum attainable A′ become larger.

Figure 9 shows plots of A′ versus $\langle\langle f_6 \rangle\rangle$ [10]. The absorbency is unambiguously determined by $\langle\langle f_6 \rangle\rangle$.

Figure 10 shows the plot of the ratio $\langle\langle f_6 \rangle\rangle / \langle\langle F \rangle\rangle$ for NaCMC from cellulose II and cellulose I. For NaCMC from cellulose II $\langle\langle f_6 \rangle\rangle / \langle\langle F \rangle\rangle$ is 0.9 ± 0.1, if $\langle\langle F \rangle\rangle$ is below 0.7. This indicates that the carboxymethylation occurs preferentially at the hydroxyl group attached to the C_6 carbon if cellulose II is utilized. On the other hand, for NaCMC from cellulose I this ratio is 0.43 ± 0.1, for $\langle\langle F \rangle\rangle$ below 0.7. These facts

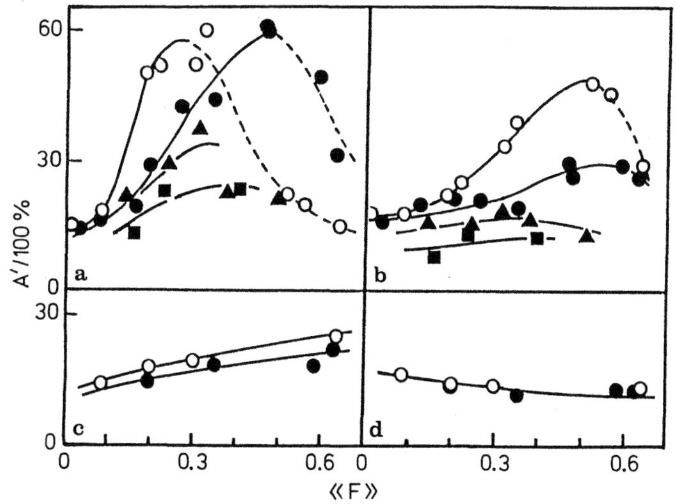

Fig. 8. Absorbency A′ of NaCMC towards various liquids as a function of total degree of substitution $\langle\langle F \rangle\rangle_{Chem}$ [10]; a: water; b: 0.9 wt% aq. NaCl; c: 0.9 wt% $CaCl_2$; d: 0.9 wt% aq. $AlCl_3$; ○: CMC from cellulose II ($M_v = 7.3 \times 10^4$); ●, CMC from cellulose I ($M_v = 21.0 \times 10^4$); ▲, CMC from cellulose I ($M_v = 9.4 \times 10^4$); ■, CMC from cellulose I ($M_v = 7.4 \times 10^4$)

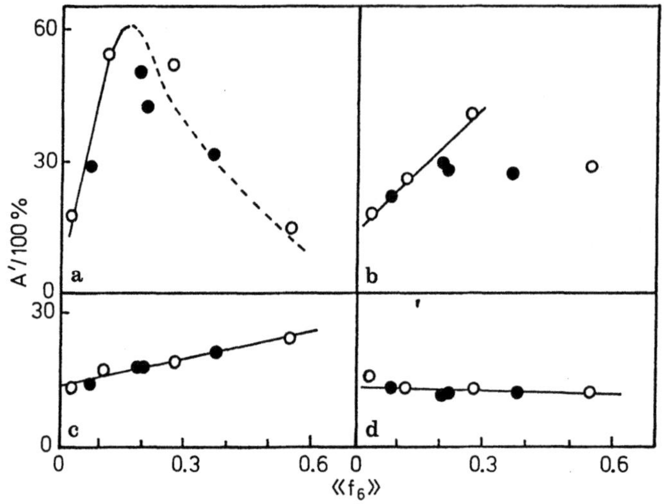

Fig. 9. Absorbency A' of NaCMC towards various liquids as a function of 《f_6》[10]; a: H_2O; b: 0.9 wt% aq. NaCl; c: 0.9 wt% aq. $CaCl_2$; d: 0.9 wt% aq. $AlCl_3$. (○, ●: as in Fig. 8.)

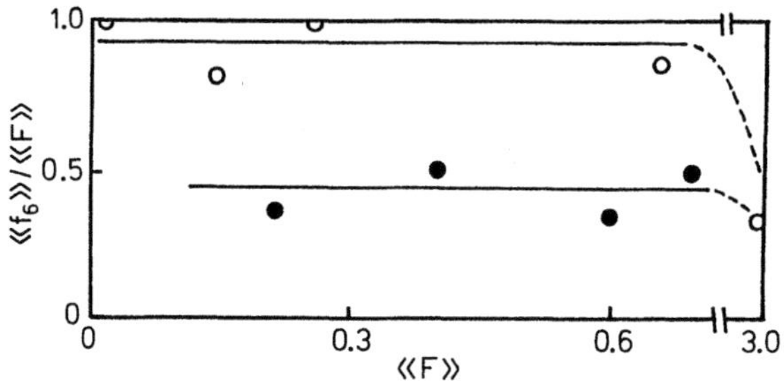

Fig. 10. Plot of 《f_6》/《F》$_{NMR}$ vs. 《F》$_{NMR}$ [10]. (○, ●: as in Fig. 8.)

show that the reactivity of the hydroxyl groups depends significantly on the cellulose sample, especially on its crystalline form.

A detailed investigation of the NMR spectrum of the C_6 carbon peak region for solid cellulose [29] revealed that hydroxyl groups at the C_6 position of cellulose II participate in two types of intramolecular hydrogen bonds (O_2—H ... O_6' and O_6—H ... O_2', the latter being much more acidic (Fig. 11)), whereas those of cellulose I do not form O_6—H ... O_2' hydrogen bonds. Moreover it was found that the electron density on the C_6 carbon of cellulose II, in both the ordered (crystalline) and amorphous regions, is higher than that for cellulose I, and the reverse is true for the electron density on the C_2 carbon. These facts suggest that the relative reactivity of the hydroxyl group at the C_6 position of cellulose II is higher than that of cellulose I, and at the C_2 position this is reversed. Thus, the preferential substitution of the hydroxyl groups at the C_6 position does not occur under the same conditions of carboxymethylation in the case of cellulose I. This is illustrated in Fig. 12. The δ^-_{I-1} and δ^-_{I-2} represent the electronegativity (EN) of the hydroxyl groups attached at C_6 and C_2 positions for

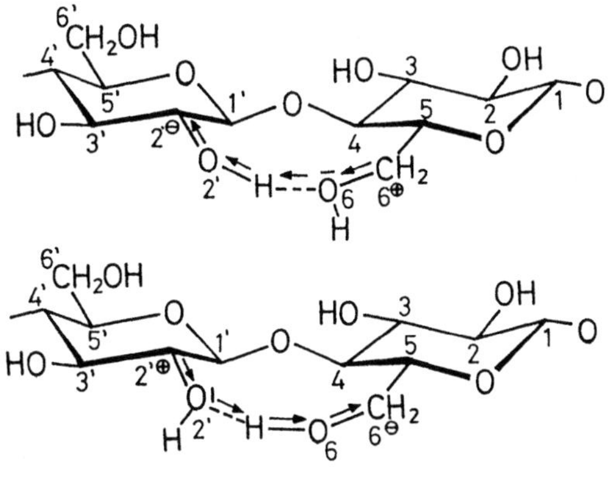

Fig. 11. Schematic representation of intramolecular hydrogen bond in cellulose molecules; → indicates the direction of movement of electron; + : NMR peak shifts to lower magnetic field; — : NMR peak shifts to higher magnetic field

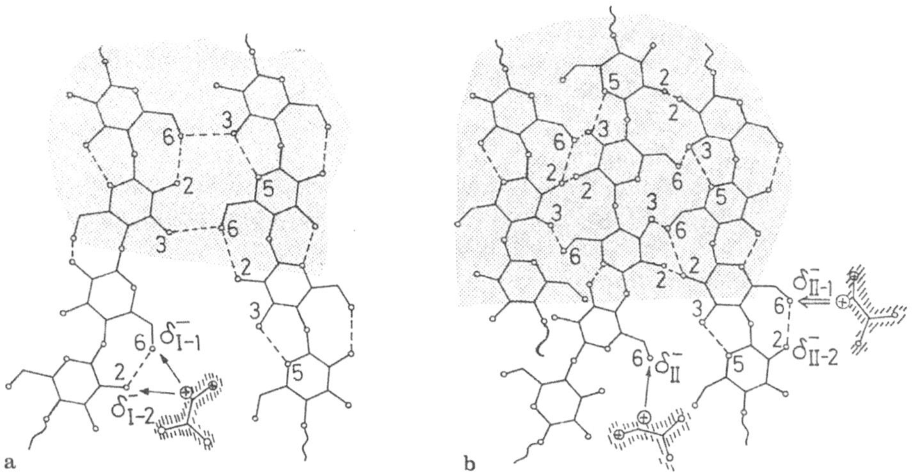

Fig. 12 a and b. Carboxymethylation towards O_2—H ... O_6' type intramolecular hydrogen bonds in cellulose I (**a**) and cellulose II (**b**) [10]: Shadowed area denotes crystalline part, hatched area cationized monochloroacetic acid molecule. (For δ^- see text)

cellulose I, and δ_{II}^-, δ_{II-1}^-, and δ_{II-2}^- the EN of the hydroxyl groups at C_6 and C_2 positions of cellulose II. The EN fall into the following order: $\delta_{II-1}^- > \delta_{II}^- > \delta_{I-1}^- > \delta_{I-2}^- > \delta_{II-2}^-$. The O_6—H ... O_2' type intramolecular hydrogen bonds for cellulose II are not shown. But, the corresponding EN on the hydroxyl group at the C_6 position is much stronger than δ_{I-1}^-. Introduction of bulky substituents at the C_6 position destroys the intermolecular hydrogen bonds and widens the distance between molecular chains, making space available in which absorbed liquid can be readily accommodated, because in original cellulose the hydroxyl groups at the C_6 position mainly govern the intermolecular hydrogen bonds.

3 Preparation of Cellulose Derivative Samples with Narrow Molecular Weight Distribution

Well-characterized polymer materials with a molecular weight distribution as narrow as possible are widely used as samples for the establishment of structure-properties-processability relationships.

New Solution Fractionation Method: Successive Solution Fractionation

The preparative fractionation methods including the solubility difference method, column fractionation, and TLC, have been extensively studied [30–36]. For example, Kamide and coworkers [31–36] concluded from computer experiments as well as actual experiments on the polystyrene/cyclohexane (or methylcyclohexane) system that successive precipitation fractionation (SPF), which is used fairly widely at present, does not give polymers with very narrow molecular weight distribution (hereafter referred to simply as monodisperse polymer) under conventional operating conditions. In contrast, the successive solution fractionation (SSF) has a good possibility of yielding monodisperse polymer samples under very easily accessible operating conditions [37]. The principal difference between SPF and SSF is schematically demonstrated in Fig. 13 [38]. In the former the polymer-rich phase is separated as the fraction and in the latter the polymer-lean phase is isolated.

Table 2 shows combinations of solvent and non-solvent suitable to fractionate CA using SSF [39–42]. For the fractionation of CTA, acetic acid and chlorinated hydrocarbons, which have a low dielectric constant ε, have been used extensively. Unfortunately, the fractionation efficiency achieved with these solvents was poor, and numerous attempts made so far have met with very limited success [43]. Judging from the easiness of two-liquid phase separation and of solvent recovery, Kamide et al. [39] chose 1-chloro-2,3-epoxy-propane (epichlorohydrine) as a solvent and heptane as a precipitant.

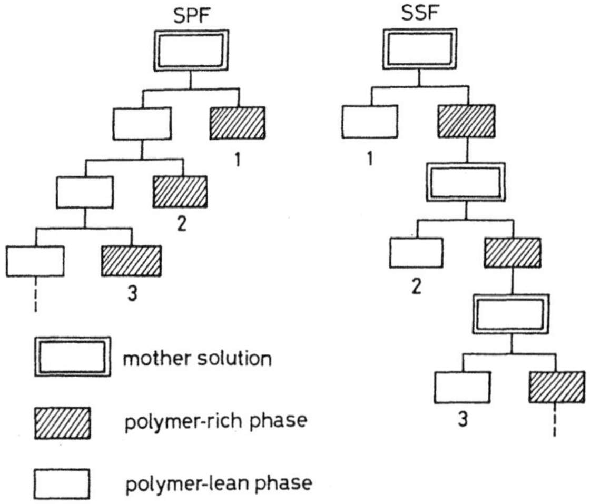

Fig. 13. Schematic representation of successive precipitation fractionation (SPF) and successive solution fractionation (SSF) [38]

Table 2. Solvents and non-solvents used to fractionate cellulose acetate (CA) with different total degree of substitution ⟪F⟫

Polymer (⟪F⟫)	Good solvent	Non-solvent	M_w/M_n	Ref.
CA(2.92)	1-chloro-2,3-epoxypropane	Hexane	1.3–1.5	[39]
CA(2.46)	Acetone	Ethanol	1.1–1.5	[40]
CA(1.75)	Acetone/water (7/3, v/v)	Water	1.2–1.5	[41]
CA(0.49)	Water	Methanol	1.1–1.5	[42]

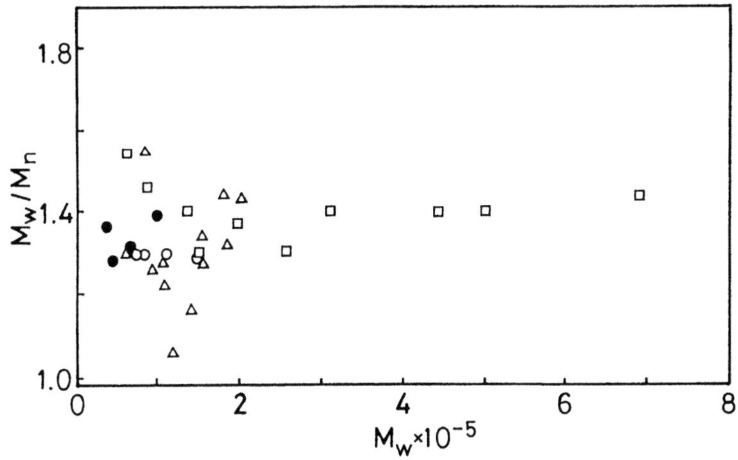

Fig. 14. Plot of M_w/M_n against M_w for CA fractions with ⟪F⟫ = 0.49 (○); 1.75 (●); 2.46 (△); and 2.92 (□), isolated by SSF method

Figure 14 shows the plots of the ratio M_w/M_n of the CA fractions prepared by the SSF method [39–42], as a function of their M_w. M_w and M_n values were determined by light scattering and membrane osmometry, respectively. Except for a few fractions of CA(2.46) and CA(2.92), M_w/M_n values of CA fractions lie between 1.2 and 1.5, independently of their M_w values. Most of the fractions of cellulose derivatives reported in the literature were prepared by the SPF method [44]; their M_w/M_n values range roughly from 1.2 to 3.7 (most of them 1.5–2.0) and moreover depend markedly on M_w. This indicates clearly that the SSF method is superior to SPF for cellulose derivatives, as it is also the case for synthetic polymers such as polystyrene [38]. A computer simulation for a quasi-ternary system carried out by Kamide and Matsuda also showed the inconditional superiority of the SSF method [34–36].

4 Light Scattering Measurements

(a) Sample Preparation

Experimental difficulties have been encountered for light scattering (LS) measurements of CA, even in good solvents, owing to contamination by gel-like materials in the solutions [40,45–48]. Kamide et al. studied extensively the nature of the prehump (i.e. gel) in GPC curves for CDA in acetone [47] and tetrahydrofuran (THF) [48] (see Fig. 15 and Table 3), and found a preparation procedure for solutions of CDA giving no prehump. It consists of choosing mild conditions of acetylation, followed by very careful hydrolysis, and treating the whole CDA polymer thus obtained with a weak acid, decomposing those CA molecules having a large M_w and containing a sulfuric acid group, sodium and calcium, and then extracting the component CDA with higher ⟪F⟫ ($A_{c,w}$ = 57–58%) (particle size, 150–450 nm in acetone, as determined by prehump I in the GPC curve) with dichloromethane.

The new technique of CDA sample preparation for light scattering measurements was successfully applied for industrial purposes, to develop a high-speed dry spinning process of CDA fiber [49].

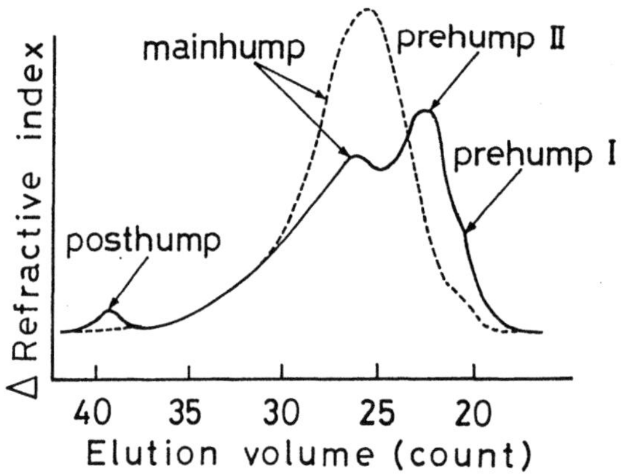

Fig. 15. GPC elution curves for cellulose diacetate (CDA) whole polymers, run through a series of columns (silica gel SI-1000, 500, 150, and 50) in acetone at 25 °C [47]. Solid line: CDA synthesized from wood pulp; dotted line: from cotton liner. GPC curves exhibit multiple peaks

Table 3. Dissolved state of cellulose diacetate in acetone at 25 °C [47]

GPC elution curve	Dissolved state
Prehump I	Molecular aggregate of cellulose acetate with higher acetyl content (57–58%); particle size, 0.15–0.45 μ.
Prehump II	Molecular aggregate of cellulose diacetate having large molecular weight, which contains sulfuric acid groups, sodium and calcium; particle size, 0.08–0.3 μ.
Mainhump	Molecularly dispersed particles.
Posthump	Degradated products of cellulose acetate?

(b) Zimm Plots

A typical Zimm plot is shown in Fig. 16. It has a non-distorted diamond shape, exhibiting no downward curvature [40].

Table 4 gives M_w data of some CA fractions determined in various solvents by LS [7]. There is excellent agreement of the M_w values in various good solvents. This fact confirms the reliability of the LS method and also indicates that the CA(0.49) and CA(2.46) polymers dissolve molecularly in those solvents, i.e. without association.

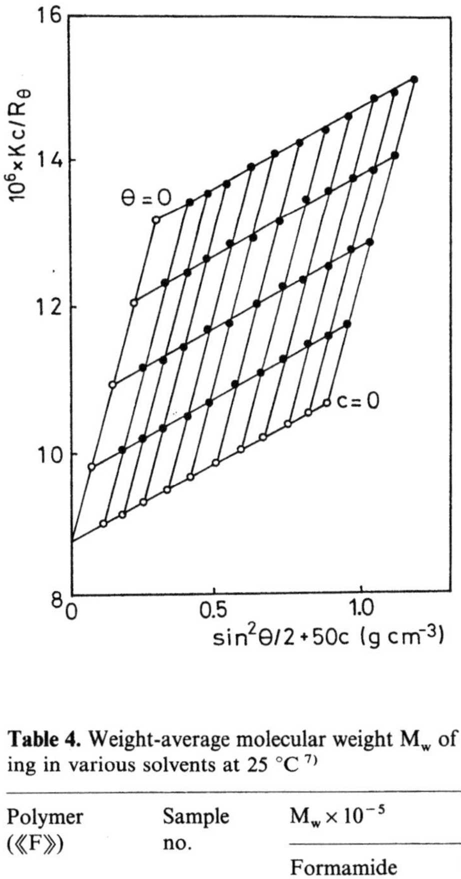

Fig. 16. Zimm plot of a cellulose diacetate fraction in acetone at 25 °C [40]

Table 4. Weight-average molecular weight M_w of cellulose acetates (CA) determined by light scattering in various solvents at 25 °C [7]

Polymer (《F》)	Sample no.	$M_w \times 10^{-5}$				
		Formamide	Water	DMAc[a]	Acetone	THF[b]
CA(0.49)	MA-5[c]	0.630	0.642	0.632	—	—
CA(2.46)	EF3-6[d]	—	—	0.73	—	0.74
	EF3-8	—	—	—	0.94	1.00
	EF3-10	—	—	1.08	1.06	1.09
	EF3-12	—	—	1.41	1.55	—
	EF3-13	—	—	1.55	1.56	—
	EF3-15	—	—	2.70	2.65	—

[a] N,N-Dimethylacetamide; [b] Tetrahydrofuran; [c] Ref. [42]; [d] Ref. [40,45]

5 Vapor Pressure Osmometry, Membrane Osmometry, and Gel Permeation Chromatography

A new vapor pressure osmometry (VPO) apparatus having a very high sensitivity has been constructed by Kamide et al. [50]. In order to match a pair of thermistors, a conventional Wheatstone bridge circuit was modified by introducing a matching resistor, which permits to detect a temperature difference of ca. 6×10^{-6} °C.

Figure 17 shows the plot of $\Delta T_s/c$ vs. c (ΔT_s = temperature difference at steady state; c = concentration of polymer) for CA in THF [51]. From the intercept of the plot at c = 0, M_n can be evaluated and compared with M_n values estimated by other methods (Table 5). The VPO method seems to yield M_n values only a few percent lower than those estimated by membrane osmometry (MO) or GPC.

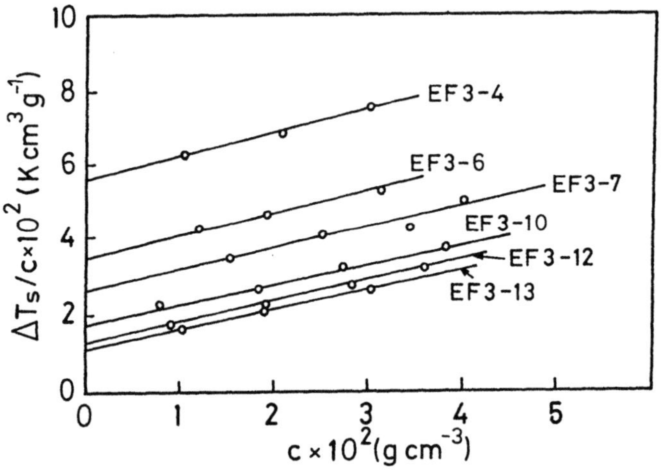

Fig. 17. Plot of $\Delta T_s/c$ against c for several cellulose diacetate fractions in THF at 25 °C [51]

Table 5. Number-average molecular weight M_n of cellulose acetate (CA) fractions with ⟪F⟫ = 2.46, determined by various methods, with tetrahydrofuran as solvent [51]

Polymer	Sample code	$M_n \times 10^{-4}$			M_w/M_n [d]
		MO [a]	GPC [b]	VPO [c]	
CA(2.46)	EF3-4	2.7	4.1	2.6	1.66
	EF3-6	4.8	5.0	4.1	1.59
	EF3-7	5.5	5.8	5.4	1.36
	EF3-10	8.7	8.8	8.1	1.25
	EF3-12	12.2	11.9	11.1	1.25
	EF3-13	12.6	12.3	12.9	1.28

[a] Membrane osmometry; [b] Gel-permeation-chromatography; [c] Vapor pressure osmometry; [d] Averaged over all data points available for each sample

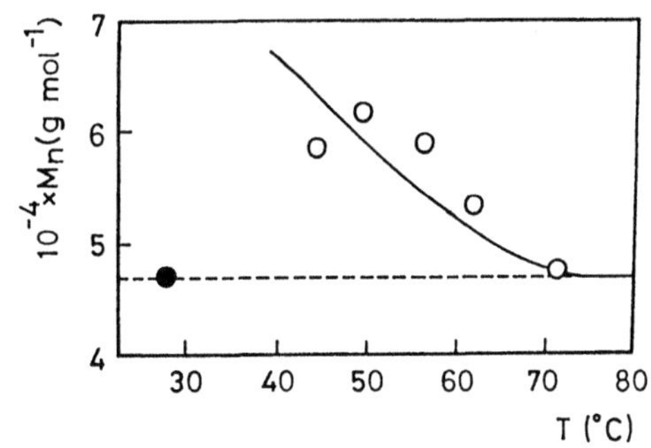

Fig. 18. Plot of apparent number-average molecular weight M_n, as determined by membrane osmometry, for a cellulose diacetate fraction in tetrachloroethane (TCE) against the temperature of measurement [40]: ○, M_n in TCE; ●, M_n in THF

The value of M_n determined by MO in tetrachloroethane (TCE) decreases with increasing temperature, approaching M_n as determined by MO in THF, as shown in Fig. 18 [40]. The CDA molecules have a considerable tendency to form aggregates in poor solvents, as is TCE below 70 °C. Ikeda and Kawaguchi [52], who overlooked this phenomenon for CDA in TCE, claimed that TCE at 56.5 °C acts as a Flory theta solvent for CDA, for which the second virial coefficient A_2 vanishes.

The CTA solution in acetone can be prepared by the cooling method: CTA dissolves if a slurry of CTA in acetone is cooled to —40 °C or below, followed by warming to room temperature [53, 54].

M_n values determined in various solvents including acetone coincide (Table 6) [39]. The CTA dissolves molecularly in acetone and the solution is very stable.

Table 6. Number average molecular weight M_n of cellulose acetate fractions with ⟪F⟫ = 2.92, determined by membrane osmometry in various solvents at 25 °C [39]

Sample code	$M_n \times 10^{-4}$			
	DMAc[a]	Acetone	TCE[b]	TCM[c]
TA2-5	5.60	5.45	5.80	5.60
TA2-10	21.9	21.1	21.2	21.7

[a] N,N-dimethylacetamide; [b] tetrachloroethane; [c] trichloromethane

CTA has been dry-spun from a solution of methylene chloride, which is corrosive and toxic to workers. Kamide et al. [54] carried out a dry-spinning of CTA in acetone, using a large-scale, commercially available spinning machine for CDA. Table 7 summarizes the tensile strength and elongation of CTA filaments prepared under various dissolving conditions [54]. Tensile strength per denier and elongation of sample ACTA-3 are almost the same as those of the filament on the market (MCTA-1). No differences in crystallinity as determined by X-ray, nor in dyeability, between samples ACTA-1 and MCTA-1 were detected.

Table 7. Preparative conditions of the CTA-acetone solutions for dry-spinning and the physical properties of CTA filaments dry-spun from acetone and methylene chloride solution [54]

Sample	Degree of substitution 《F》	Solvent	Disolving conditions	Physical properties of filaments				
				Denier (d)	Tensile strength (g/d)		Elongation (%)	
					Dry	Wet	Dry	Wet
ACTA-1	2.70	Acetone	Cooled down three times successively	123.9	1.26	0.79	33.4	37.9
ACTA-2	2.70	Acetone	Once cooled down and then condensed	72.1	1.12	0.72	27.3	33.5
ACTA-3	2.75	Acetone	Cooled down three times successively	74.7	1.13	0.75	30.4	35.2
ACTA-4	2.90	Acetone	Cooled down three times successively	189.5	1.19	0.78	18.5	23.4
MCTA-1 (commertial fibers)	2.90	Methylene chloride	—	74.7	1.27	0.84	36.0	39.0

6 Critical Phenomena of Solutions of Cellulose Derivatives

Figure 19 depicts plots of the temperature of the observed cloud point vs. the weight fraction w_2 of CA(2.46) for each solution [55,56]. The lower critical solution temperature, LCST, was determined as the minimum temperature of each cloud-point curve. Some CD-solvent systems show the existence of an upper critical solution temperature, UCST, together with LCST (Fig. 20) [57].

Figure 21 shows a schematic representation of the mechanism of pore formation in the solvent casting process [58]. Depending on the initial polymer concentration, v_p^0, the polymer-rich phase or the polymer-lean phase separates initially from the solution. If v_p^0 is smaller than the critical polymer concentration, v_p^c, the polymer-rich phase separates as small particles (primary particles). These particles grow into secondary particles. The mechanism disclosed here is of general validity and provided a powerful method for controlling the average pore size and the pore size distribution of polymeric membranes, including cellulose and CA membranes [58].

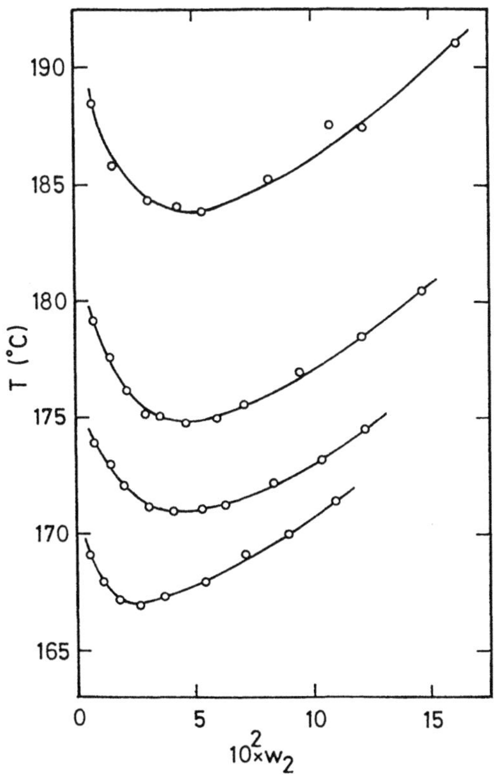

Fig. 19.

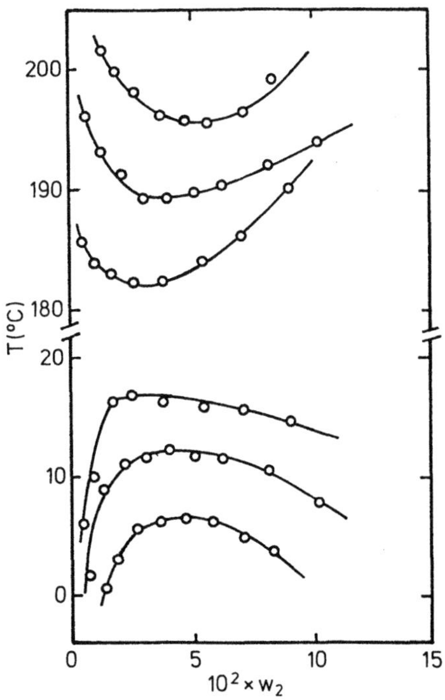

Fig. 20.

Fig. 19. Plots of the temperature of the cloud point versus the weight fraction of cellulose diacetate fraction-acetone systems [56]

Fig. 20. Phase diagram for cellulose diacetate fraction-2-butanone systems, showing LCST and UCST [57]

Table 8. Concentration dependent parameters p_1 and p_2 in the χ parameter, Flory's theta temperature θ, and the entropy parameter ψ of cellulose diacetate, determined by different methods [61]

Solvent	UCSP or LCSP	Methods									
		Kamide-Matsuda [59]				Koningsveld et al. [62]				Shultz-Flory [63]	
		p_1	p_2	θ	ψ	p_1	p_2	θ	ψ	θ	ψ
acetone	LCSP	0.96	−21.4	433.1	−1.03	0.203	−11.78	429.7	−1.01	427.1	−0.78
methylethylketone	UCSP	−0.67	2.52	312.2	0.47	−0.850	4.40	313.5	0.46	311.4	0.35
	LCSP	−0.96	7.38	432.0	−0.69	−1.170	9.56	430.7	−0.68	434.6	−0.51

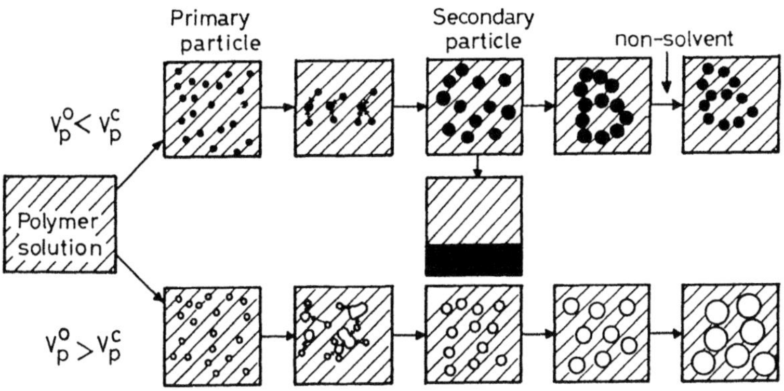

Fig. 21. Schematic representation of the formation of pores in the casting process [58]

Kamide and Matsuda [59] proposed a theory to determine the concentration dependence coefficients p_1 and p_2 of the polymer-solvent interaction parameter χ $\{\equiv \chi_0(1 + p_1 v_p + p_2 v_p^2)$; χ_0 = concentration independent coefficient; v_p = volume fraction of polymer$\}$, Flory's theta temperature θ, and the entropy parameter ψ for multicomponent polymer-single solvent systems, using the critical solution temperature T_c, v_p^c, the critical χ_0 parameter χ_0^c, and the weight- and z-average relative molar volume ratios of polymer to solvent, X_w and X_z. The method was successfully applied to polystyrene and polyethylene in various solvents [60]. Table 8 summarizes the results on CA(2.46) [61]. The θ temperature can be determined independently of the method used. But p_1 and p_2 of CA(2.46) solutions determined by the Kamide and Matsuda [58] and Koningsveld et al. [61] methods do not coincide, because of the low experimental accuracy of v_p^c. The absolute value of ψ determined by the above two methods is larger than that obtained by the Shultz-Flory [63] method, in which the concentration dependence of χ is ignored, contrarily to the case of polystyrene-solvent systems [60]. In CA(2.46)-solutions solvation phenomena occur, as described later, and the temperature dependence of the solvation possibly affects p_1 and p_2.

7 Thermodynamic Properties of Solutions of Cellulose and its Derivatives

The linear expansion coefficient α_s ($\equiv \langle S^2 \rangle^{1/2}$ and $\langle S^2 \rangle_0^{1/2}$ are the radii of gyration in the perturbed and unperturbed state, respectively) is determined from the penetration function Ψ which is defined by [64]

$$\Psi \equiv \bar{z} h_0(\bar{z}) = 0.746 \times 10^{-25} A_2 M^2 / \langle S^2 \rangle^{3/2} \tag{7}$$

where

$$\bar{z} \simeq z/\alpha_s^3 \tag{8}$$

$$z = (3/2\pi)^{3/2} B A^{-3} M^{1/2} \tag{9}$$

$$A = (\langle R^2 \rangle_0 / M)^{1/2} \tag{10}$$

$$B = \beta' / m_0^2 \tag{11}$$

A and B are short- and long-range interaction parameters (A is also termed the unperturbed chain dimension), $\langle R^2 \rangle_0^{1/2}$ is the mean-square end-to-end distance of the chain in the unperturbed state, β' the binary cluster integral, and m_0 is the molecular weight of a segment. According to the Kurata-Fukatsu-Sotobayashi-Yamakawa theory [65], $h_0(\bar{z})$ is related to $\bar{z}$ by

$$\bar{z} h_0(\bar{z}) = (1/5.047)\{1 - (1 + 0.683\bar{z})^{-7.39}\} \tag{12}$$

The value of z can be determined from experimental data on A_2, M, and $\langle S^2 \rangle_0^{1/2}$ by using Eqs. (7) and (12). α_s is related to z through [66]

$$\alpha_s^3 - 1 = 1.78z \tag{13}$$

Combination of Eq. (8) and (13) furnishes a method for estimating α_s from $\bar{z}$. The coefficient (1.78) in Eq. (13) differs slightly depending on the theory, but this variation does not afford a significant change in α_s in the range $\alpha_s^3 < 2$ and this is the case of almost all CD solutions [2].

Figure 22 shows α_s of CA solutions as a function of M_w [7]. The α_s values are smaller than 1.18 for $M_w \leqq 7 \times 10^5$, except for those of a few CA(0.49) fractions with molecular

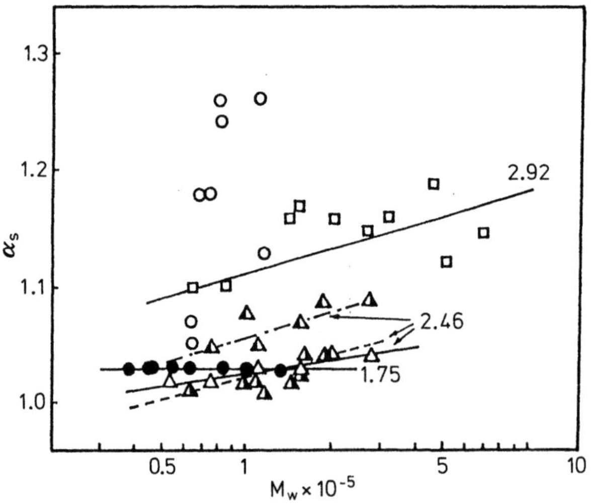

Fig. 22. Linear expansion coefficient α_s of cellulose acetate (CA)-solvent systems plotted as a function of M_w [7]. The lines are determined by the least-square method. Numbers on the lines denote the total degree of substitution ⟪F⟫ of CA. ○: CA(0.49)-DMAc; ●: CA(1.75)-DMAc; △: CA(2.46)-DMAc; ▲: CA(2.46)-acetone; ◭: CA(2.46)-THF; □: CA(2.92)-DMAc

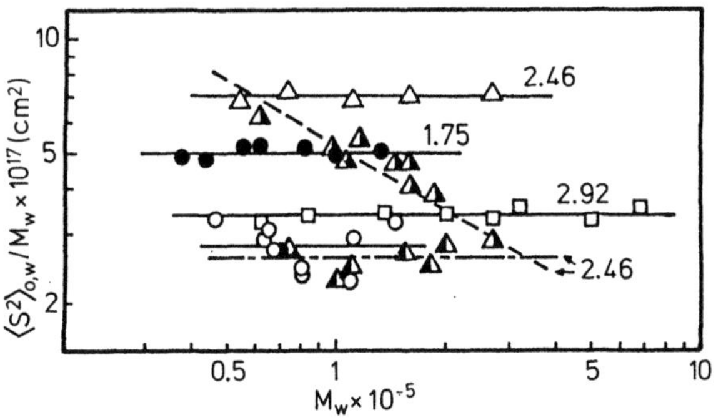

Fig. 23. Molecular weight dependence of the mean-square weight average radius of gyration in the unperturbed state relative to M_w ($\langle S^2\rangle_{0,w}^{1/2}/M_w$) for cellulose acetate in various solvents [7]. The lines are determined by the least-square method. The symbols are the same as those in the legend of Fig. 22

weights from 8×10^4 to 1×10^5. The α_s values calculated according to the wormlike touching beads model are smaller than those from the pearl necklace model [67–69]. The excluded volume effect in CA solutions is evidently small, irrespective of the model used.

Figure 23 depicts the plots of $\langle S^2\rangle_{0,w}/M_w$ versus M_w [7]. Except for the CA(2.46)-acetone system, $\langle S^2\rangle_{0,w}/M_w$ of the CA solutions is independent of M_w over the range of M_w from 3×10^4 to 7×10^5, indicating that CA molecules with a wide variety of ⟪F⟫ dissolved in N,N-dimethylacetamide (DMAc), as well as CA(2.46) molecules in THF are Gaussian chains. In contrast, CA(2.46) molecules in acetone behave as non-Gaussian chains. Kamide and Miyazaki [2], analyzing literature data, found that $\langle S^2\rangle_{0,w}/M_w$ of solutions of cellulose and CD such as cellulose tricaproate (CTCp), methyl cellulose (MC), NaCMC, hydroxyethylcellulose (HEC), ethyl hydroxyethylcellulose (EHEC), and sodium cellulose xanthate (NaCX) decreases with increasing M_w.

The existence of a hydrogen bond between the O-acetyl group in the CA molecule and halogenated hydrocarbons [70, 71], as well as the interaction between the O-acetyl group and aniline or acidic solvents have been suggested by Marsden and Urpuhart [72] on the basis of infrared spectroscopic observations. NMR spectroscopy is also a powerful tool for studying the interaction of functional groups in CD molecules with a solvent.

Kamide et al. [7, 73] measured the chemical shift of the O-acetyl and hydroxyl groups in CA(0.49), CA(1.75), CA(2.46), and CA(2.92) whole polymers in various solvents by ^{1}H-NMR. The data are summarized in Table 9. Solvents with a high dielectric constant strongly interact with both O-acetyl and hydroxyl groups of the CA molecules (see also Figs. 24, 25) [73].

Assuming an incompressible part of the solution, Passynsky derived an equation relating the adiabatic compressibility β, observed for the solution, to the number n of grams of solvating solvent per g of polymer [73]:

$$n = (1 - \beta/\beta_s)(100\varrho - c)/c \quad (14)$$

Table 9. NMR peaks of O-acetyl (O-Ac) and hydroxyl (OH) protons of cellulose acetate (CA) in various solvents [7)]

Solvent	Dielectric constant ε	NMR peaks (ppm)							
		CA(0.49)[a]		CA(1.75)		CA(2.46)		CA(2.92)	
		O-Ac	OH	O-Ac	OH	O-Ac	OH	O-Ac	OH
Formamide	111	2.17	3.72	—	—	—	—	—	—
Deuterium oxide (D_2O)	78	2.19	4.70	—	—	—	—	—	—
TFA[b]	39.5	2.22	—	—	—	2.24	—	2.23	—
DMAc[c]	37.8	2.21	3.87	2.23	3.93	2.23	3.90	2.15	3.84
Acetone	20.7	—	—	—	—	2.06	2.85	2.04	2.75
THF[d]	8.2	—	—	—	—	2.00	2.51	—	—
Pyridine	—	2.10	4.92	—	—	2.09	4.83	2.08	4.80
DCM[e]	7.77	—	—	—	—	2.05	1.62	2.03	1.64
TCE[f]	7.29	—	—	—	—	—	—	2.00	1.70
DMF[g]	—	—	—	—	—	—	—	2.07	3.50

[a] numbers in parentheses denote the total degree of substitution; [b] trifluoroacetic acid; [c] N,N-dimethylacetamide; [d] tetrahydrofuran; [e] dichloromethane; [f] trichloroethane; [g] N,N-dimethylformamide.

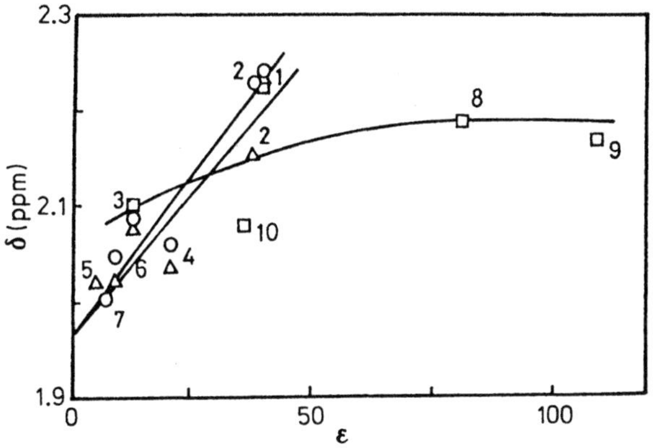

Fig. 24. Plot of the weight-average methylproton chemical shifts of cellulose acetate (CA) molecules versus the dielectric constant ε of the solvents [73]; rectangles: CA(0.49); circles: CA(2.46); triangles: CA(2.92)

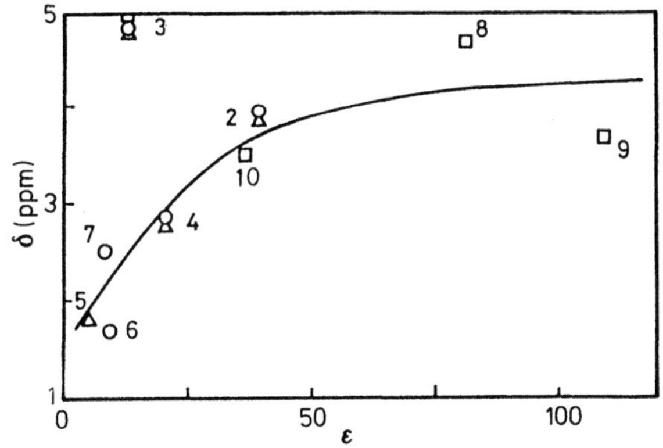

Fig. 25. Plot of the weight-average hydroxyl-proton chemical shifts of cellulose acetate (CA) versus the dielectric constant ε of the solvents [73]; see legend of Fig. 24

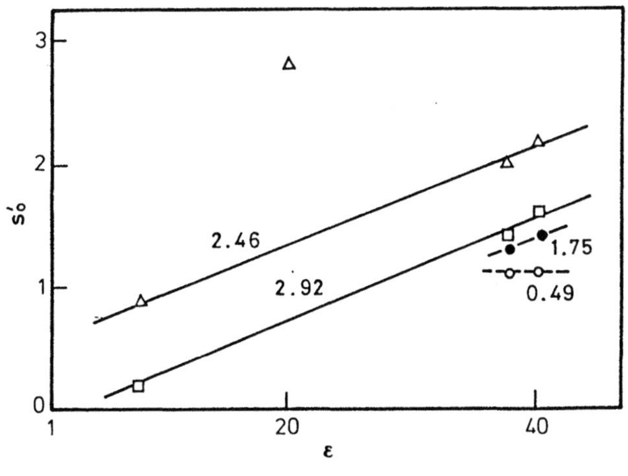

Fig. 26. Relation between the number of solvating solvent molecules per glucose ring at infinite dilution and the dielectric constant ε, for cellulose acetate (CA)-solvent systems at 25 °C [7]. The numbers on the lines denote the total degree of substitution ⟪F⟫; ○: CA(0.49); ●: CA(1.75); △: CA(2.46); □: CA(2.92)

where β_s denotes β of the solvent, ϱ the density of the solution; β_s and β can be determined experimentally from sound velocity measurements. The number n can be converted to the number of solvating solvent molecules per repeating unit at infinite dilution, s_0', using Eq. (15):

$$s_0' = \lim_{c \to 0} (m_p/m_s)\, n \tag{15}$$

Here m_p and m_s are the molecular weight of the repeating unit of the polymer and of the solvent, respectively.

Moore et al. determined s_0' of solutions of CN(2.33) [75], CA(2.25), CA(2.99) [76], and ethyl cellulose [76], using an ultrasonic interferometer. Recently, Kamide and

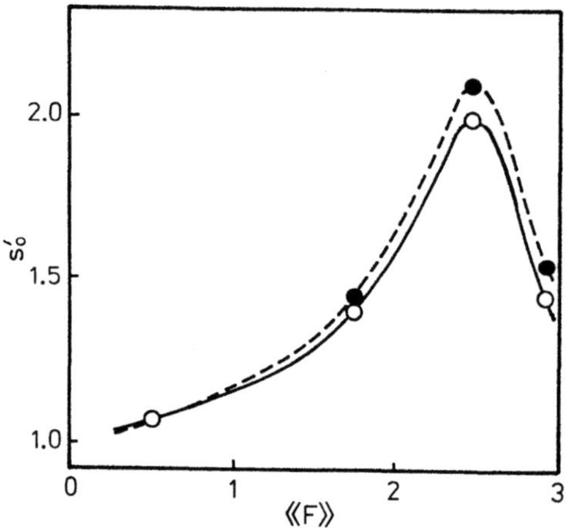

Fig. 27. The effect of the total degree of substitution 《F》 on the number of solvating solvent molecules per glucose ring at infinite dilution s_0', for cellulose acetate (CA)-DMAc and CA-DMSO at 25 °C [7]; ○: DMAc; ●: DMSO

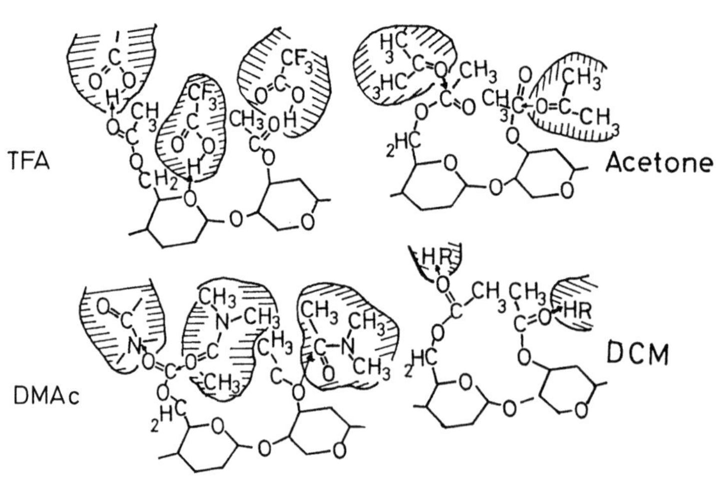

Fig. 28. Possible molecular models of the interaction between cellulose acetate with 《F》 = 2.92 and various solvents [72]

Saito [7] measured the sound velocity in well-characterized CA solutions with a Pierce type ultrasonic interferometer with high accuracy and determined s_0'. Figure 26 shows the relation between s_0' of CA whole polymer solutions and ε of the solvent at 25 °C [7]. While the dependence of s_0' on ε differs depending on ⟪F⟫, s_0' except for CA(2.46)-acetone increases with increasing polarity of the solvent in a similar manner as the chemical shifts of the O-acetyl and hydroxyl groups. In Fig. 27, the effects of ⟪F⟫ on s_0' for CA-DMAc and CA-dimethylsulfoxide (DMSO) solutions at 25 °C are shown. In both systems, s_0' has a maximum at ⟪F⟫ $\simeq$ 2.5 [7].

Thus, we have been able to estimate the strength of the interaction between solvent and CD molecules with the aid of NMR, and the number of solvating solvent molecules with the acoustic method.

Figure 28 shows the possible molecular models of the interactions of CTA in various solvents [73].

8 Hydrodynamic Properties (I)

a) Mark-Houwink-Sakurada Equation

The molecular weight dependence of [η] can be expressed by the semi-empirical Mark-Houwink-Sakurada (MHS) equation,

$$[\eta] = K_m M_w^a \tag{16}$$

Table 10 summarizes the parameters K_m and a of CD-solutions. With a few exceptions (CA(2.46)-DMAc, CN-acetone [78,79], cellulose tricarbanilate (CTC)-acetone and -dioxane [80,81], HEC-water [82], and EC-water [83]), a is <0.8, as for most synthetic polymer solutions [91].

If M_n is used instead of M_w in Eq. (16), the exponent a occasionally exceeds 0.8. For example, all literature data on K_m and a for CTA-dichloromethane (DCM), and trichloromethane (TCM) indicate an appreciable departure from those by Kamide et al. [92] (Fig. 29). In first approximation, the difference can be explained (with the exception of the results of Nair et al. [93]) by a broad molecular weight distribution of the samples used. In particular, a systematic molecular weight dependence of the molecular weight distribution of the samples will yield incorrectly high a values [92].

The exponent a in the MHS equation can be divided into four parts

$$a = 0.5 + a_\Phi + a_1 + 1.5a_2 \tag{17}$$

with

$$a_1 \equiv 3\mathrm{d} \ln \alpha_s/\mathrm{d} \ln M \tag{18}$$

$$a_2 \equiv \mathrm{d} \ln (\langle S^2\rangle_0/M)/\mathrm{d} \ln M \tag{19}$$

and

$$a_\Phi \equiv \mathrm{d} \ln \Phi/\mathrm{d} \ln M \tag{20}$$

Table 10. K_m and exponent a in the Mark-Houwink-Sakurada equation, and exponent a_Φ, for solutions of cellulose and its derivatives

Polymer (《F》)	Solvent	$K_m \times 10^2$ / $cm^3\ g^{-1}$	a	a_Φ	Ref.
Cellulose	Cadoxen	0.0338	0.77	0.304	84)
	FeTNa[a]	0.0531	0.775	0.429	85)
Cellulose acetate(2.92)	TFA[b]	3.96	0.70_6	—	39)
	DMAc[c]	2.64	0.75	0.106	39)
	Acetone	2.89	0.75_5	—	39)
	TCM[d]	4.54	0.64_6	—	39)
	DCM[e]	2.47	0.70_4	—	39)
Cellulose acetate(2.46)	DMAc	1.34	0.82	0.23	45)
	Acetone	0.13	0.616	0.716	40)
	THF[f]	0.51_3	0.68_8	0.105	40)
Cellulose acetate(1.75)	DMAc	9.58	0.65	0.12	41)
Cellulose acetate(0.49)	Formamide	20.9	0.60	—	42)
	Water	20.9	0.60	—	42)
	DMSO[g]	17.1	0.61	—	42)
	DMAc	19.1	0.60	0.103	42)
Cellulose nitrate(2.91)	Acetone	0.76	0.903	0.379	78, 79)
Cellulose nitrate(2.55)	Acetone	0.48	0.916	0.274	78, 79)
Cellulose tricaproate	DMF[h]	0.245	0.5	0.377	86)
	1-Cl—N[i]	0.17	0.51	0.377	86)
Cellulose tricarbanilate	Acetone	1.43×10^{-3}	0.91	0.21	80, 81)
	Dioxane	8.13×10^{-4}	0.97	0.462	80, 81)
Methylcellulose(2.3)	Water	0.316	0.55	0.464	87)
Hydroxyethyl cellulose(0.88)	Water	8.62×10^{-3}	0.87	0.608	82)
Ethyl hydroxy-ethyl cellulose ($-OC_2H_4$; 0.84; $-OC_2H_5$; 0.56)	Water	0.037	0.8	0.688	83)
Sodium cellulose xantate(0.78)	1 M NaOH	2.47	0.679	0.568	88, 89)
Sodium carboxy methyl cellulose(1)	NaCl ($I \to \infty$)	0.19	0.60	0.192	90)

[a] Iron sodium tartrate; [b] trifluoroacetic acid; [c] N,N-dimethylacetamide; [d] trichloromethane; [e] dichloromethane; [f] tetrahydrofuran; [g] dimethylsulphoxide; [h] N,N-dimethylformamide; [i] 1-chloronaphthalene

Here Φ is Flory's viscosity parameter, defined as

$$\Phi \equiv [\eta]\, M/(6\langle S^2 \rangle)^{3/2} \tag{21}$$

Figure 30 a–c demonstrates the dependence of a, a_Φ, and a_2 of CA-DMAc solutions on 《F》[7, 39–42, 45]. For these systems, both the exponent a in the MHS equation and a_Φ reach a maximum at 《F》 $\simeq$ 2.5, but a_2 is almost independent of 《F》. Inspection of Fig. 30 indicates that the main factor contributing to a is a_Φ.

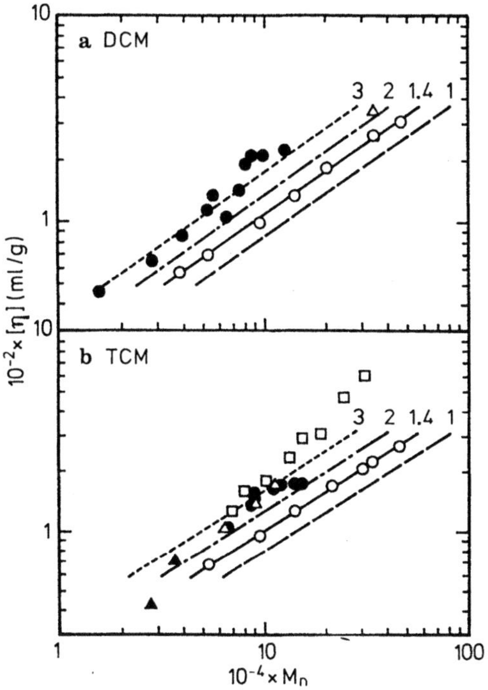

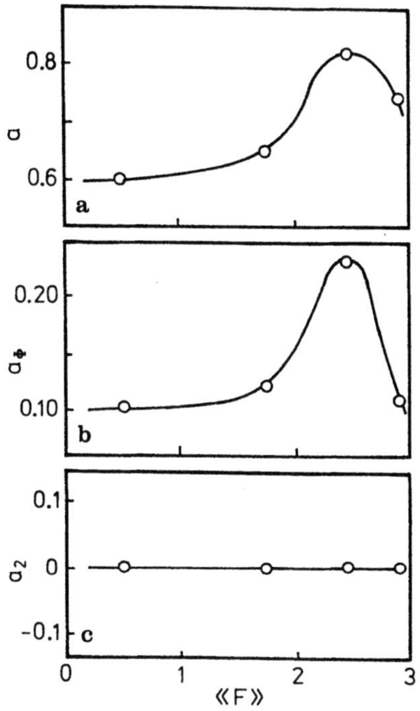

Fig. 29.

Fig. 30.

Fig. 29. Double logarithmic plot of limiting viscosity number $[\eta]$ versus the number-average molecular weight M_n for cellulose triacetate solutions [92]. a) dichloromethane (DCM); ○: Kamide et al. [92]; ●: Dymarchuk et al. [94], △: Shakhparonov et al. [95]; b) trichloromethane (TCM); ○: Kamide et al. [82]; △: Staudinger and Eicher [96]; □: Sharples and Major [97]; ▲: Howard and Parikh [98]; ●: Nair et atl. [93]

Fig. 30. The dependence of the exponent a (Eq. (16)), a_Φ (Eq. (20)), and a_2 (Eq. (19)) on the total degree of substitution ⟪F⟫ for cellulose acetate-DMAc solutions [7]

b) Molecular Weight Dependence of the Sedimentation Constant s_0

Figure 31 a and b show the plots of s_0 as a function of M_w, for CA(2.46)-acetone and CA(2.92)-DMAc, respectively (open circles) [99]. The following relationships were obtained by Ishida et al. [99] using the least-square method (s_0 in seconds) at 25 °C:
CA(2.46)-acetone

$$s_0 = 1.02 \times 10^{-14} M_w^{0.390} \quad (22)$$

CA(2.92)-DMAc

$$s_0 = 1.83 \times 10^{-14} M_w^{0.242} \quad (23)$$

Note that these equations were established for well-characterized fractions with narrow molecular weight distribution. Eq. (22) also fits data of Singer [100] and of Holmes and Smith [101].

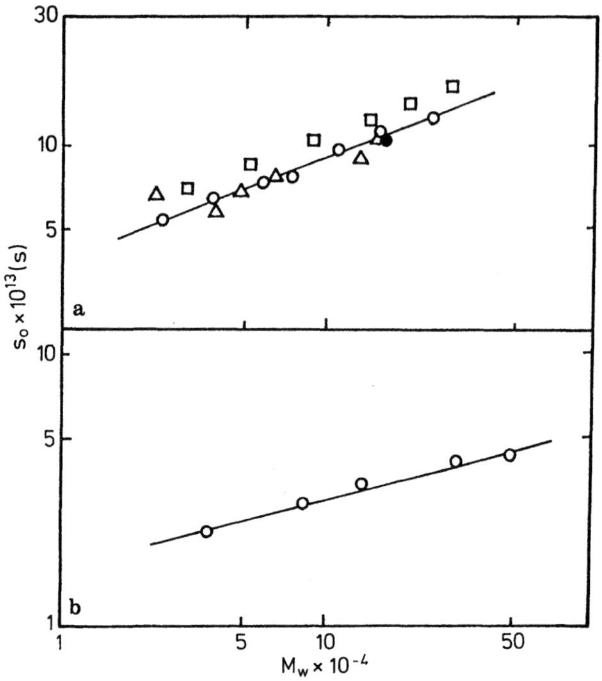

Fig. 31. Molecular weight dependence of the sedimentation constant s_0: a) cellulose acetate (CA) with ⟪F⟫ = 2.46 in acetone at 25 °C; Ishida et al. [99] (○); Singer [100] (●); Holmes and Smith [101] (△); Golubev and Frenkel [102] (□); b) cellulose acetate with ⟪F⟫ = 2.92 in DMAc at 25 °C [99]

If all data hitherto obtained [99–102] (except those of Golubev and Frenkel [102]) are taken into account, the following relationship between M_w and s_0 for CA(2.46) in acetone is obtained:

$$s_0 = 1.70 \times 10^{-14} M_w^{0.345} \tag{24}$$

c) Temperature dependence of $[\eta]$

It is well known that $[\eta]$ of most CD solutions decreases significantly with increasing temperature [80, 82, 83, 86, 90, 103–109]. Even in theta solvents, this effect has been observed for cellulose tricaprylate [104] and CTCp [86]. This finding is in sharp contrast with what is known for vinyl polymers.

The negative temperature coefficient of $[\eta]$ has long been considered characteristic of CD and has been discussed by several researchers. Mandelkern and Flory [104], Flory et al. [105], Krigbaum and Sperling [86], and Moore and Edge [108] ascribed it to the rapid decrease of A (see Eq. (10)) with increasing temperature. Kamide et al. [109] and Shanbhag [80] disagreed with this interpretation and preferred to consider volume effects rather than skeletal effects of the molecular chains. All these interpretations were made as early as in the 1950's and 1960's without direct measurements of the molecular dimensions [110].

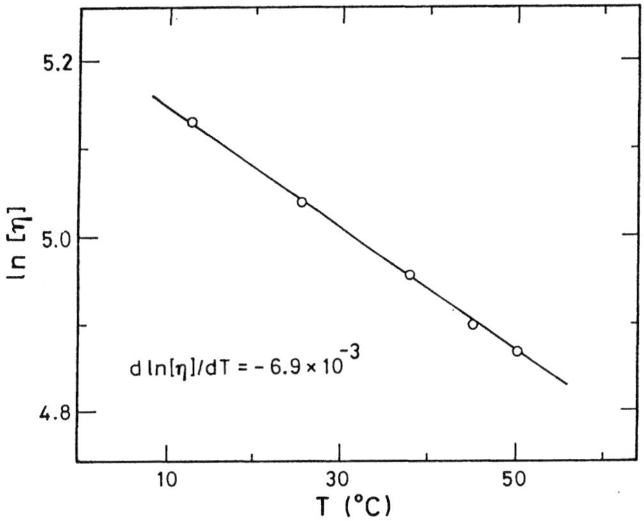

Fig. 32. Temperature dependence of ln [η] for a cellulose diacetate fraction in acetone [110]

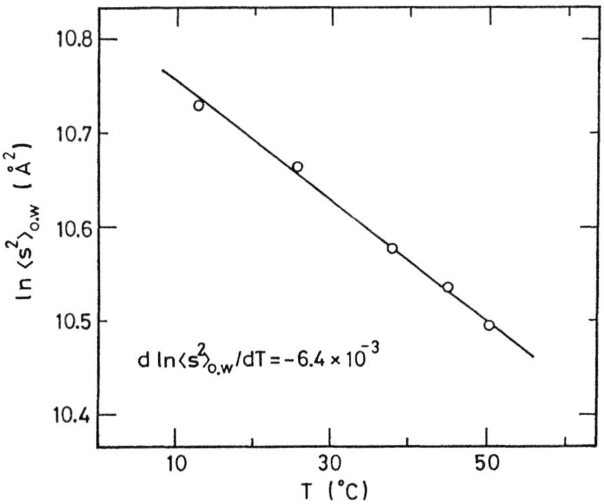

Fig. 33. Temperature dependence of ln $\langle S^2 \rangle_{0,w}$ for a cellulose diacetate fraction in acetone [110]

In general, the temperature dependence of [η] can be written as

$$\begin{aligned} d \ln [\eta]/dT = {} & d \ln \Phi/dT \\ & + 1.5 d \ln \langle S^2 \rangle_{0,w}/dT \\ & + 3 d \ln \alpha_s/dT \end{aligned} \tag{25}$$

Suzuki et al. [110] measured [η] and $\langle S^2 \rangle_z^{1/2}$ of CA(2.46)-acetone in the temperature range 10–50 °C and estimated $\langle S^2 \rangle_0$ using α_s obtained by a penetration function method. Figs. 32 and 33 show the temperature dependence of [η] and $\langle S^2 \rangle_0$ for a CA(2.46)-acetone solution [110]. The following values were obtained for the above solution:

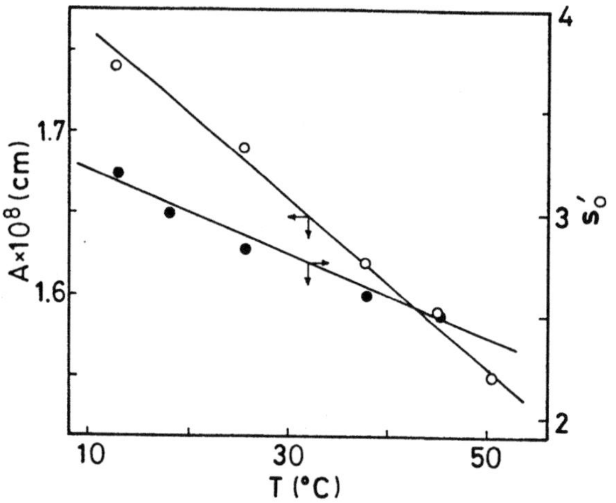

Fig. 34. Temperature dependence of the unperturbed chain dimension A, estimated by method 2B (Eq. (32)), for a cellulose diacetate (CDA) fraction in acetone (open circles), and of the number of the solvating solvent molecules at infinite dilution s'_0 of a CDA whole polymer in acetone (black circles) [7]

$$d \ln [\eta]/dT = -6.9 \times 10^{-3} \tag{26}$$

$$d \ln \Phi/dT = (2.8 \pm 0.5) \times 10^{-3} \tag{27}$$

$$d \ln \langle S^2 \rangle_{0,w}/dT = -6.4 \times 10^{-3} \tag{28}$$

$$d \ln \alpha_s/dT = 0 \tag{29}$$

It can be deduced therefore, that the rapid decrease in the unperturbed chain dimension with increasing temperature is the dominant cause of the large negative value of d ln [η]/dT. The effect of the excluded volume is negligible.

Figure 34 shows the temperature dependence of A of a CA(2.46) fraction with $M_w = 9.4 \times 10^4$ in acetone, and s'_0 of the CA(2.46) whole polymer in acetone [7]. Here A was estimated by method 2B (Eq. (32), see Section 10). A as well as s'_0 decreases monotonically with rising temperature. Thus it is evident that d[η]/dT can be interpreted in terms of ds'_0/dT.

9 Hydrodynamic Properties (II): Draining Effect

Two conflicting theoretical views concerning the flexibility of polymer chains and the role of the volume effect and the draining effect on [η] are discussed in the literature: polymer chains of typical flexibility such as vinyl polymer chains, and a large value of [η] can be interpreted in terms of the excluded volume effect (view point A); polymer chains are semi- or inflexible and their large unperturbed chain dimension is mainly responsible for a large [η] (view point B). The former has its foundation on the "two parameter theory" [111]. Untill 1977 these inconsistencies constituted one of the most outstanding problems yet unsolved in the science of polymer solutions.

Kamide et al. [2,39–42,45] intended to solve this problem in a very rigorous and systematic manner and to show that view point B is preferable.

Cellulose Molecules in Solution are Partially Free Draining Chains

Evidence 1

Kurata and Stockmayer [112] predicted in an excellent review that Φ should remain essentially constant at its asymptotic value $\Phi_0(\infty)$. However, their prediction was wrong, because a_Φ, which is a more rigorous criterion of the draining effect, is always distinctly positive, as shown at the last column in Table 10. As reported by Kamide and Miyazaki [2], in cellulose and CD solutions Φ can never be treated as a constant value, over a wide range of molecular weight assessed.

Evidence 2

The draining parameter X is defined for the "pearl necklace model" (Fig. 35) as,

$$X \equiv (3/2\pi)^{1/2}\,(d/a')\,N'^{1/2} \tag{30}$$

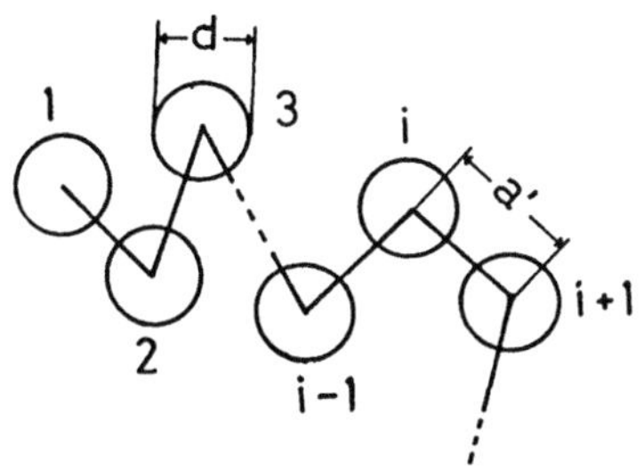

Fig. 35. Schematic representation of the pearl necklace model

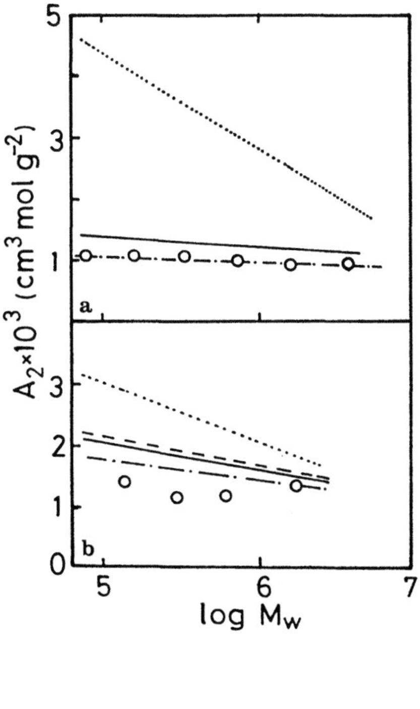

Fig. 36a and b. Second virial coefficient A_2 of **a)** cellulose nitrate (CN) (Nc = 13.9%) and **b)** CN (Nc = 12.9%), in acetone [78,79]; ○: experimental data [2]. Lines are calculated by using the penetration function ψ from short and long range interaction parameters A and B, which are estimated by methods 2C (full line), 2D (broken line), 2E (dotted line), and 2G (chain line), together with experimental $\langle S^2\rangle_w$ data

Table 11. The draining parameter X evaluated by various methods for solutions of cellulose and its derivatives [2, 39–42, 45]

Polymer	Degree of substitution ⟪F⟫	Solvent[a]	X					
			1A	1B	1C	1D	1E	1F
Cellulose	0	Cadoxen	2.2–0.2 (5.4)[b]	—	1.6	0.9	0.3	—
	0	FeTNa	0.3–1.6 (1.0)	—	—	0.4	—	—
Cellulose acetate	2.92	DMAc	1.4–2.9 (2.1)	0.3	—	—	0.3–1.0 (0.7)	—
	2.46	DMAc	0.4–0.7 (0.6)	—	—	—	—	—
		Acetone	0.5–4.0 (1.4)	1.4	—	—	0.3–0.4 (0.3)	—
		THF	4.6–8.5 (6.6)	—	—	—	—	—
	2.45	THF	0.7	—	—	—	—	—
	2.45	TFE	0.3–0.8 (0.46)	—	—	—	—	—
	2.43	Acetone	0.04–0.07 (0.06)	—	—	—	—	—
	2.33	Acetone	—	0.74	0.74	—	—	—
	1.75	DMAc	0.8–1.3 (1.1)	—	—	—	—	—
	0.49	DMAc	4.2–10.5 (7.3)	—	—	—	—	—
Cellulose nitrate	2.87	Acetone	0.3–2.4 (1.2)	—	—	0.7	—	—
	2.85	Acetone	—	0.66	1.13	—	—	—
	2.71	Acetone	2.6–30 (9.9)	—	—	0.2	—	—
	2.66	Ethyl acetate	—	0.73	—	—	—	—
Cellulose nitrate	2.57	Acetone	—	—	—	—	0.13–0.29	—
		Ethyl acetate	—	—	—	—	0.17–0.62	—
	2.39	Acetone	0.7–1.5 (1.0)	—	—	1.8	—	—
Cellulose tricaproate	3	DMF	0.3–0.6 (0.4)	—	—	0.8	—	—
		1-Cl—N	0.2–0.7 (0.4)	—	—	1.1	—	—
		Dioxane/water	0.63	—	—	—	—	—
Cellulose tricarbanilate	3	Acetone	0.4–1.8 (1.1)	—	—	0.9	—	—
		Cyclohexanone	0.5–1.6 (1.0)	—	—	4.0	—	—
		Dioxane	0.3–1.6 (0.8)	—	—	—	—	—
Methyl cellulose	1.53	Water	0.3–0.8 (0.6)	—	—	0.8	0.6–2.7	0.21
	1.39	Acetone	—	2.5	2.0	—	0.09–1.3	—
Ethyl cellulose	2.73	Acetone	—	1.1	1.1	—	—	—

Table 11. (continued)

Polymer	Degree of substitution ⟪F⟫	Solvent[a]	X					
			1A	1B	1C	1D	1E	1F
Sodium carboxy methylcellulose	1	NaCl (I → ∞)	4.3–∞ (8.1)	0.2	—	2.6	—	2.1
Hydroxy ethyl cellulose	0.88	Water	0.4–3.6 (1.9)	—	—	—	—	—
		Cardoxen	0.5–4.2 (2.9)	—	—	0.13	—	—
Ethyl hydroxy ethylcellulose	OC_2H_4; 0.84 OC_2H_5; 0.56	Water	0.5–4.3 (2.0)	0.12	0.1	—	—	—
Sodium cellulose xanthate	0.78	1 M NaOH	0.1–0.2 (0.14)	0.03	0.39	0.01	0.31–∞	—

[a] solvents as in Table 10; [b] average value in parentheses

a′ is the length of a link, d the hydrodynamic diameter of a segment, and N′ is the number of segments.

X can be evaluated by using the following methods [2]: method 1A (Φ and α_s), 1B (exponent a in the MHS eq. and exponent a_s in the relationship of the sedimentation coefficient at infinite dilution vs. molecular weight), 1C (a and the exponent a_d in the relationship of the diffusion coefficient vs. molecular weight), 1D (a, the molecular weight dependence of the radius of gyration λ and a_2), 1E (the concentration dependence of the sedimentation coefficient s and [η]), 1F (a, a_s, and λ). Considering the experimental accuracy, we cannot evaluate the exact value of X, but the different methods always give $X \leqq 2$ with some exceptions, as indicated in Table 11 and this is considerably lower than the value for vinyl-type polymers. It appears reasonable to assume that the partially free draining effect on [η] cannot be ignored in the case of cellulose and CD [2].

Evidence 3

From the values of A, B, and X, together with the experimental $\langle S^2 \rangle_w^{1/2}$ data, A_2 of cellulose trinitrate (CTN) was calculated [2]. Figure 36 shows the smooth curves of the molecular weight dependence of A_2 calculated as the full (method 2C, Eq. (33)), dotted (method 2E, Eq. (36)), broken (method 2D, Eq. (34)), and chain (method 2G, Eq. (38)) lines. The experimental data points are also included in this figure. There is a considerable disagreement between the experimental points and the theoretical curve by method 2E, in which the draining effect is neglected. However, the above disagreement should be remarkably improved by considering the draining effect and a non-Gaussian nature, in the case of CTN-acetone.

Evidence 4

If we assume a molecular chain model, a′ and N′ (Eq. (30)) can be separately evaluated and then d can be estimated from X data alone [45].

Another approach for evaluating d if X is unknown uses first-oder perturbation theory for the volume effect [113]:

$$(\alpha_s^2 - 1)/\{1 - (\theta/T)\} = (134/105)\,(6/\pi)^{1/2}\,(d/a')^3\,N'^{1/2} \tag{31}$$

Table 12. Evaluation of hydrodynamical segment diameter d of cellulose diacetate and polystyrene [45]

Polymer	Solvent	Sample No.	$M_w \times 10^{-4}$	X_E [a]	$d \times 10^8$/cm Eq. (30) $X = X_E$	$d \times 10^8$/cm Eq. (30) $X = 20$	$d \times 10^8$/cm Eq. (31)
Cellulose	Acetone	EF3-9	9.6	0.95	33.5	712	24.6
diacetate	DMAc [b]	EF3-13	15.6	0.71	37.0	1047	—
Polystyrene	Cyclohexane	HB2-3	320	33.1	9.3	5.6	7.3

[a] X value estimated from experimental data; [b] N,N-dimethylacetamide

Substitution of the experimental values for α_s and θ which is estimated by the Shultz-Flory method as listed in Table 8, into Eq. (31) permits to determine d. The d values determined on the basis of the statistical chain segment model by these two methods (Eq. (30) and Eq. (31)) are quite consistent (Table 12). The use of $X = 20$ (very weak draining effect) in Eq. (30) yields an unreasonably large d value, nearly 30 times larger than the d value (24.6 Å) obtained by the thermodynamic method. This indicates that the draining effect cannot be neglected.

10 Unperturbed Chain Dimension

Various methods have been proposed for the estimation of A of CD solutions

Method 2B [2]

$$\langle S^2\rangle_0^{1/2} = \langle S^2\rangle^{1/2}/\alpha_s \tag{32}$$

α_s is estimated with Ψ from A_2, M, and $\langle S^2\rangle^{1/2}$ data.

Method 2C [114] *(Baumann plot)*

$$\langle S^2\rangle^{3/2}/M^{3/2} = A^3/6^{3/2} + (1/4\pi^{3/2})\, BM^{1/2} \tag{33}$$

Method 2D [2] *(Baumann-Kamide-Miyazaki plot)*

$$\langle S^2\rangle^{3/2}/M^{3(1+a_2)/2} = K_0^{3/2} + (1/4\pi^{3/2})\, BM^{(1-3a_2)/2} \tag{34}$$

with

$$K_0 = \Phi_0(\infty)\, A^3 \tag{35}$$

Method 2E [115] *(Stockmayer-Fixman plot)*

$$[\eta]/M^{1/2} = K_0 + 2(3/2\pi)^{3/2}\, \Phi_0(\infty)\, BM^{1/2} \tag{36}$$

Method 2F [116] *(Kamide-Moore plot)*

$$\begin{aligned} &-\log K_m + \log\,[1 + 2\{a - 1/2)^{-1} - 2\}^{-1}] \\ &= -\log K_0 + (a - 1/2)\log M_0 \end{aligned} \tag{37}$$

M_0 is a parameter depending on the molecular weight range ($M_1 - M_2$) for which the MHS equation applies (approximately, M_0 is replaced by $(M_1M_2)^{1/2}$).

Method 2G [2] *(Kamide-Miyazaki(I) plot)*

$$[\eta]/M^{1/2+a_\Phi+3a_2/2} = 6^{3/2}K_\Phi K_0^{3/2} + 0{,}66 K_\Phi BM^{(1-3a_2)/2} \tag{38}$$

where

$$K_\Phi = \Phi/M_\Phi^a \tag{39}$$

Method 2I [117] *(Cowie-Bywater plot)*

$$\xi/\eta_0 = P_0(\infty)\,\alpha_s^{-0.348} AM^{1/2}(1 + 0.201BA^{-3}M^{1/2} - \ldots) \tag{40}$$

with

$$\xi = (1 - \bar{v}_p\varrho)\, M/(s_0N_A) \tag{41}$$

$$P \equiv (\xi/\eta_0)/(6^{1/2}\langle S^2\rangle^{1/2}) = K_pM_p^a \tag{42}$$

$P_0(\infty)$ is P in the unperturbed and nondraining state; $\bar{v}_p$, ϱ, η_0 are the molar volume of the polymer, viscosity of the solvent, and density of the solution, respectively; N_A is the Avogadro number.

Method 2J [118] *(Kamide-Miyazaki (II) plot)*

$$\xi/\eta_0 = 6K_pK_0^{1/2}M_p^{1/2 + a_2/2 + a}(1 + 0.0144BK_0^{-3/2}M^{1/2 - 3a_2/2}) \tag{43}$$

Method 2K [119] *(Kamide-Saito (I) plot)*

$$\log K_m + \log(1 - 3a_2)/2(1 - a + a_\Phi) + (3/2)\log 6 + K_\Phi$$
$$= -(3/2)\log K_0 + \left(a - \frac{1}{2} - a_\Phi - 3a_2/2\right)\log M_0 \tag{44}$$

Method 2L [120] *(Kamide-Saito (II) plot)*

$$-\log K_\xi + \log(1 - 3a_2)/(2(1 - a_\xi + a_p - a_2)) + \frac{1}{2}\log 6 + \log K_p$$
$$= -\frac{1}{2}\log K_0 + \left(a_\xi - a_p - \frac{1}{2}a_2 - \frac{1}{2}\right)\log M_0 \tag{45}$$

where K_ξ and a_ξ are parameters in the relationship:

$$\xi/\eta_0 = K_\xi M_\xi^a \tag{46}$$

Table 13 summarizes the A values of CD solutions estimated using method 2B–2L [2,7,119,120]. The A values estimated by methods 2E, 2F, and 2I agree well, but are about 20–40% lower than those obtained by methods 2B, 2C, 2D, 2G, 2J, 2K, and 2L, due to neglect of the draining effect and non-Gaussian nature. The method based on the two-parameter theory assuming $a_2 = a_\Phi = a_p = 0$ is not applicable to CD solutions due to the existence of the draining effect and also of the non-Gaussian nature.

Table 13. The unperturbed chain dimension A estimated by various methods for solutions of cellulose and its derivatives [7,119]

Polymer (《F》)	Solvent	A × 10^8 (cm)									
		Thermodynamic approach			Hydrodynamic approach						
		2B Eq. (32)	2C Eq. (33)	2D Eq. (34)	2E Eq. (36)	2F Eq. (37)	2G Eq. (39)	2I Eq. (40)	2J Eq. (43)	2K Eq. (44)	2L Eq. (45)
Cellulose	Cadoxen	1.53	1.83	—	1.21	1.20	1.57	1.43 (s_0[a]) 1.18 (D_0[b])	1.92	1.56	1.70
	FeTNa	1.96	2.31	2.27	1.31	1.29	2.17	—	—	2.07	—
CA[c](0.49)	Formamide	3.03	—	—	1.19	1.22	—	—	—	—	—
	Water	2.51	—	—	1.17	1.22	—	—	—	—	—
	DMAc	1.28	1.38	1.38	1.12	1.14	1.40	—	—	1.45	—
CA(1.75)	DMAc	1.73	1.73	1.73	1.12	1.16	1.79	—	—	1.73	—
CA(2.46)	DMAc	2.05	1.98	1.98	0.79	0.92	1.90	—	—	1.85	—
	Acetone	1.66[d]	2.14	1.77[e]	1.10	1.12	1.75[e]	1.11	1.72	1.76	1.80
	THF	1.23	1.24	1.24	0.99	1.00	1.23	—	—	1.26	—
CA(2.92)	DMAc	1.43	1.49	1.49	0.89	1.00	1.47	0.57	1.63	1.46	1.68
CN[f](2.55)	Acetone	1.84	1.84	1.85	0.52	0.74	1.91	—	—	1.96	—
CN(2.91)	Acetone	2.41	2.43	—	0.79	1.02	2.45	—	—	2.50	—
CTC_p[g]	DMF	—	2.14	1.94	0.96	0.92	1.99	—	—	2.00	—
	1-Cl—N	1.75	2.14	1.94	0.90	0.85	1.82	—	—	1.80	—
CTC[h]	Acetone	1.43	1.43	—	0.56	0.56	1.25	—	—	1.23	—
	Dioxane	1.88	1.85	—	0.55	0.47	1.83	—	—	1.82	—
	Cyclohexane	1.36	1.43	—	0.67	0.67	1.25	1.96	1.48	1.65	1.54
MC[i](2.3)	Water	2.05	2.56	2.34	1.23	1.25	2.31	—	—	2.34	—
NaCMC[j](1)	NaCl ($I \rightarrow \infty$)	1.42	1.50	1.47	1.26	1.25	1.44	—	—	1.44	—
HEC[k](0.88)	Water	2.10	2.40	2.25	0.93	0.96	2.26	—	—	2.26	—
EHEC[l] (OC_2H_4, 0.84; OC_2H_5, 0.56)	Water	2.10	2.41	2.37	1.24	1.22	2.32	1.08 (s_0)	2.16	2.40	2.67
NaCX[m](0.78)	1 M NaOH	3.30	4.04	3.86	1.03	1.03	3.96	0.73	3.63	3.97	4.31

[a] Sedimentation constant; [b] diffusion constant; [c] cellulose acetate; [d] averaged value over all fractions; [e] A value at $M_m = 1 \times 10^5$; [f] cellulose nitrate; [g] cellulose tricaproate; [h] cellulose tricarbanilate; [i] methylcellulose; [j] sodium carboxymethyl cellulose; [k] hydroxyethyl cellulose; [l] ethyl hydroxyethyl cellulose; [m] sodium cellulose xanthate

11 Persistence Length Evaluated using the "Wormlike Chain Model"

The conformation parameter σ ($\equiv A/A_f$, where A_f is A of a hypothetical chain with free internal rotation) for cellulose and its derivatives lies between 2.8–7.5 [2,119,120] and the characteristic ratio C_∞ ($\equiv A_\infty^2 M_b/l^2$, where A_∞ is the asymptotic value of A at infinite molecular weight, M_b is the mean molecular weight per skeletal bond, and l the mean bond length) is in the range 19–115. These unexpectedly large values of σ and C_∞ suggest that the molecules of cellulose and its derivatives behave as semi-flexible or even inflexible chains. For inflexible polymers, analysis of dilute solution properties by the pearl necklace model becomes theoretically inadequate. Thus, the applicability of this model to cellulose and its derivatives in solution should be carefully examined.

Equilibrium rigidity is conventionally expressed in terms of the persistence length q based on the "wormlike chain model" (Fig. 37). Three methods for estimating q have been adopted [67].

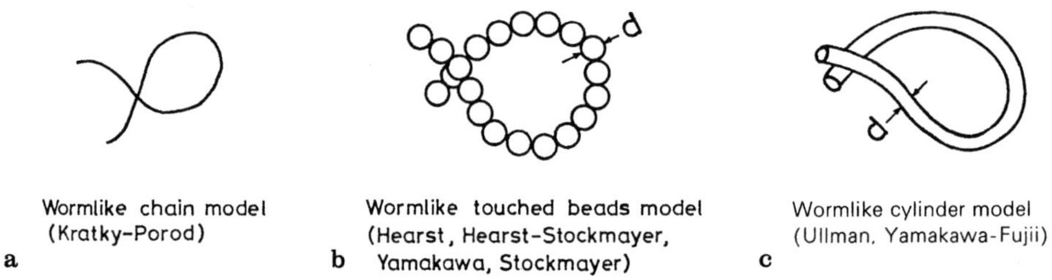

Fig. 37a–c. Schematic representation of various wormlike chain models [67]; d is the diameter of the touched bead and cylinder models

Method I Using Light Scattering Data (Benoit-Doty's Method)

Using the Benoit-Doty expression for $\langle S^2\rangle_0$ of a monodisperse wormlike chain [121], the relationship between q and $\langle S^2\rangle_z$ of polymers having a Schulz-Zimm type molecular weight distribution is derived as follows [67];

$$\langle S^2\rangle_{0,z} = q^2\left[\frac{M_w(h'+2)}{3qM_L(h'+1)} - 1 + \frac{2qM_L}{M_w} - \frac{2q^2M_L^2(h'+1)}{h'M_w^2}\left\{1 - \left(\frac{qM_L(h'+1)}{qM_L(h'+1)+M_w}\right)^{h'}\right\}\right] \quad (47)$$

where

$$h'^{-1} = M_w/M_n - 1 \quad (48)$$

$$M_L = M_w/L \quad (49)$$

$$L = N'l \quad (50)$$

Here L is the contour length.

Substituting the experimental data, such as $\langle S^2\rangle_z$, M_w, and M_n and l into Eqs. (47) to (50), we obtain q, which is denoted as q_{BD}.

By incorporating the excluded-volume effect into the wormlike touched-bead chain (Fig. 37b), Yamakawa-Stockmayer (YS) derived Ψ as a function of $\bar{z}$ in the form [122]

$$\Psi = (\langle S^2\rangle_0/\langle S^2\rangle_{0,\infty})^{-3/2}\, \bar{z}h(\bar{z}) \tag{51}$$

where

$$\langle S^2\rangle_{0,\infty} = qM/(3M_L) \tag{52}$$

$$h(\bar{z}) = 1 - Q(L, q, d)\, \bar{z} + \ldots \tag{53}$$

Here Q(L, q, d) is a function of L, q, and d of a bead. Yamakawa and Stockmayer also derived the relation between α_s and z [122]

$$\alpha_s^2 = 1 + (67/70)\, K(L, q)\, z + \ldots \tag{54}$$

where K(L, q) is a function of L and q. The following closed expression of the Fixman type was derived [67,68]

$$\alpha_s^3 - 1 = (3/2)\, 67K(L, q)/70z \tag{55}$$

Saito et al. [67,68] derived a closed form of $h(\bar{z})$ as

$$\bar{z}h(\bar{z}) = \frac{1 - \{1 + (3.903/2.865)\, Q\bar{z}^{-0.478}\}}{(1.828/2.865)\, Q} \tag{56}$$

q and α_s are determined numerically from Eqs. (55) and (56), using A_2, M_w, $\langle S^2\rangle_z$, and M_w. The resulting values are denoted by q^0_{BD} and $\alpha^w_{s,3}$.

Method II Using Light Scattering Data (Gaussian Chain Approximation Method)

For indefinitly large $M/(qM_L)$, an unperturbed polymeric chain may be regarded as Gaussian. For this chain, q is related to the unperturbed chain dimension A_∞ at infinite M_w by

$$q = M_L A_\infty^2/2 \tag{57}$$

The q value estimated by Eq. (57) is denoted as q^0_{CL}.

Method Using the Limiting Viscosity Number (Yamakawa and Fujii's Method)

Neglecting the draining term in the Kirkwood-Riseman integral equation [123], Yamakawa and Fujii [124] derived an expression of $[\eta]$ for unperturbed wormlike cylinders (Fig. 37c), which reads

$$[\eta] = \Phi(2q/M_L)^{3/2}\, M^{1/2} \tag{58}$$

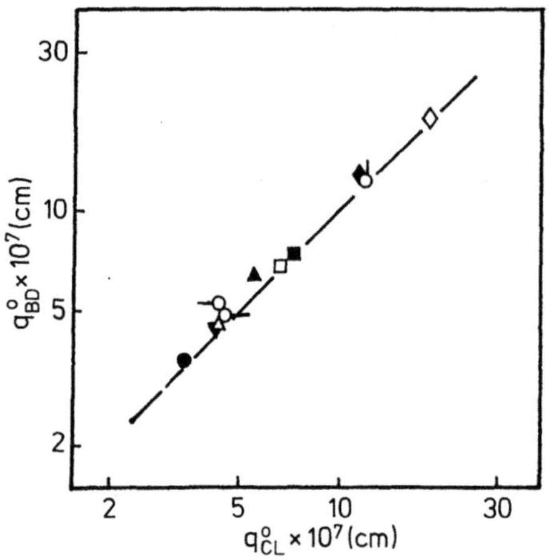

Fig. 38. Correlation between the unperturbed Benoit-Doty persistence length q^0_{BD} and the persistence length q^0_{CL} [67]; △: cellulose/cadoxen; ▲: cellulose/FeTNa; ○: CA(2.92)/DMAc; ♁: CA(2.46)/DMAc; –○: CA(2.46)/acetone; ○–: CA(2.46)/THF; ●: CA(0.49)/DMAc; ◇: CN(2.91)/acetone; ◆: CN(2.55)/acetone; □: HEC/water; ■: EHEC/water; ▽: NaCMC/aq. NaCl. The solid line represents the relation $q^0_{BD} = q^0_{CL}$

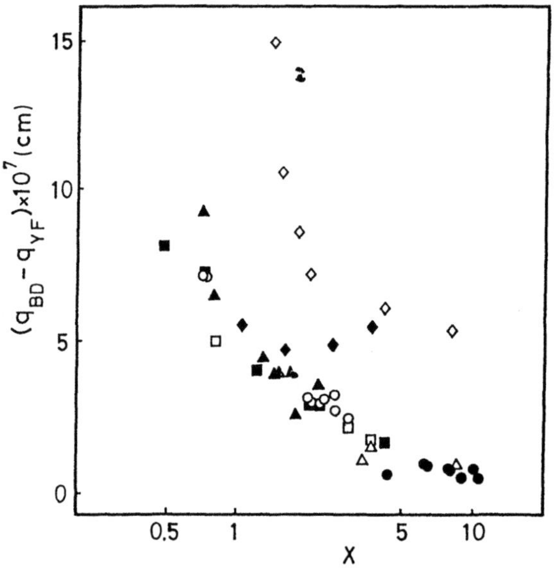

Fig. 39. The difference between the Benoit-Doty persistence length q_{BD} and the Yamakawa-Fujii persistence length q_{YF}, $q_{BD} - q_{YF}$, plotted versus the draining parameter X for cellulose and its derivatives in solution [67]. The symbols in the figure are the same as in Fig. 38

Saito determined q, referred to as q_{YF}, using experimental data, [η], M_w and d for each polymer samples [67].

Figure 38 shows the relationship between q^0_{BD} and q^0_{CL} for the highest molecular weight sample of each cellulose derivative [67]; q^0_{BD} agrees well with q^0_{CL} for any of the cellulose derivatives investigated, implying that A in the pearl necklace model varies with q in the wormlike chain model for these highest molecular weight samples.

The q_{BD} values do not agree closely with q_{YF} for any of the cellulose derivatives examined. Figure 39 shows the difference $q_{BD} - q_{YF}$ plotted against X, as determined by method 1A [67, 68]. For all cases $q_{BD} > q_{YF}$ and $q_{BD} - q_{YF}$ has a negative correlation with X (correlation coefficient $\gamma = 0.60$). We can conclude that $q_{BD} > q_{YF}$ due to the significant role played by the free draining effect.

Table 14. Persistence length of stiff and flexible polymers estimated by various methods [67)]

Polymer	Solvent	$\frac{q}{nm}$	Method
Poly(vinylpyridiniumbutyl bromide)	5×10^4 M Aq sodium chloride	70.4	V[a]
Poly(methylacrylic acid)	5×10^{-4} M Aq sodium chloride	62.5	V
Deoxyribonucleic acid (DNA)	0.79 M Aq sodium chloride	52.0	V
	0.2 M Aq sodium chloride	56.6	V
Poly(hexyl isocyanate)	n-Hexane	42.0	LS[b], V, s[c]
Poly(n-butyl isocyanate)	Tetrachloromethane	37.5	V
Polymethylmethacrylate	Benzene	1.4	SAXS[d]
Polyethylene	Bis-2-ethylhexyl adipate	0.96[e]	LS
Poly(vinyl acetate)	Heptane/3-methyl-2-butanone	1.6^5	LS
Polystyrene	Cyclohexane	0.9–1.0	SAXS

[a] viscometry; [b] light scattering; [c] sedimentation; [d] small angle X-ray scattering; [e] estimated from unperturbed chain dimensions reported in the literature

The q_{BD}^0 values of cellulose and its derivatives lie between 3 and 25 nm and are larger than those of typical vinyltype polymer ($\sim$1 nm), but markedly smaller than those of typical stiff chain polymers, such as DNA (Table 14) [67)]. Thus, the chains of cellulose and its derivatives can be considered as semi-flexible. It may be concluded that both the pearl-necklace chain and the wormlike chain models are adequately applicable to these polymers.

12 Effect of Total Degree of Substitution and Nature of the Solvent on the Chain Flexibility of Cellulose Derivatives

The value of A of CA dissolved in a common solvent depends significantly on $\langle\langle F \rangle\rangle$, as shown in Fig. 40 [7)]. In the figure, the numbers denote ε of the solvent and the solid line represents the relationship between A in DMAc and $\langle\langle F \rangle\rangle$, showing a maximum at $\langle\langle F \rangle\rangle \simeq 2.5$, as in the case of s_0' (Fig. 27). The A value in DMAc extrapolated to $\langle\langle F \rangle\rangle = 0$ (i.e. cellulose) (filled circle) is about 1.3 Å and the corresponding σ is 2.25. The open rectangles represent A values extrapolated to $\varepsilon = 1$, as given in Fig. 41 [7)]. In a hypothetical non-polar solvent ($\varepsilon = 1$), A is expected to decrease monotonically with $\langle\langle F \rangle\rangle$ (chain line, Fig. 40), approaching the value of the freely rotating cellulose chain (broken line) in the region of $\langle\langle F \rangle\rangle \leqq 0.8$. An intercept of the A vs. $\langle\langle F \rangle\rangle$ line for $\varepsilon = 1$ (chain line) at $\langle\langle F \rangle\rangle = 0$ (i.e., $\lim_{\substack{\varepsilon \to 1 \\ \langle\langle F \rangle\rangle \to 0}} A$) should be the A value of cellulose molecules dissolved in a hypohetical nonpolar solvents. $\lim_{\substack{\varepsilon \to 1 \\ \langle\langle F \rangle\rangle \to 0}} A$ represents the flexibility of the unsubstituted cellulose chain, without any solvation. Since all known solvents for cellulose, including cadoxen and iron sodium tartrate, are highly polar or electrolytic, the solution properties of cellulose in less polar solvents have never been investigated. Figure 40 indicates that cellulose, dissolved in a hypothetical nonpolar solvent, would behave almost as a freely rotating chain; in other words,

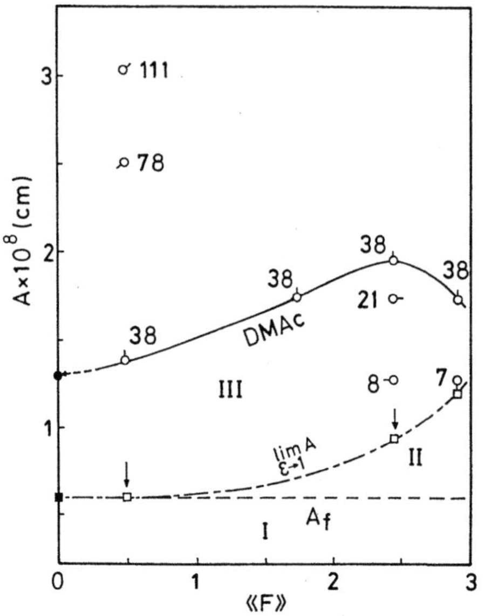

Fig. 40. Plot of the unperturbed chain dimension A against the total degree of substitution for cellulose acetate (CA)-solvent systems. Solid line: CA-DMAc; chain line: asymptotic A at the limit of the dielectric constant ε = 1; broken line: A of cellulose at the free rotational state [7]; ●: asymptotic A value at the limit of ε = 1; ■: asymptotic A value at the limit of ε = 1 and «F» = 0; ♂: formamide; ⚲: water; ♁: DMAc; ○-: acetone; -○: THF; ♀: TCE

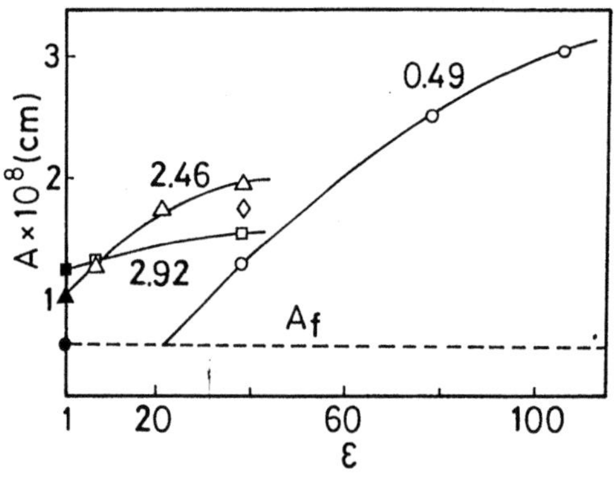

Fig. 41. Dependence of the most probable unperturbed chain dimension A on the dielectric constant ε of the solvent for cellulose acetate (CA)-solvent systems [7]. Broken line: the unperturbed chain dimension of a hypothetical cellulose with free internal rotation. Open marks: A_m values estimated from experimental data; closed marks: asymptotic A_m values at the limit of ε = 1. Circles: CA(0.49); diamond: CA(1.75), triangles: CA(2.46); rectangle: CA(2.92)

cellulose is intrinsically very flexible and the low degree of flexibility of the cellulose chain, deduced from the physical properties of cellulose in solution and in the solid state, is caused either by the solvation or by intra- or inter-molecular hydrogen bonding.

CA may be considered to have the following structural factors controlling the rigidity of the molecule: an interdependent rotational potential, which controls the unperturbed dimension in solution, by (1) the steric interactions between neighboring pyranose rings with and without substituent groups, (2) the intramolecular hydrogen bonds

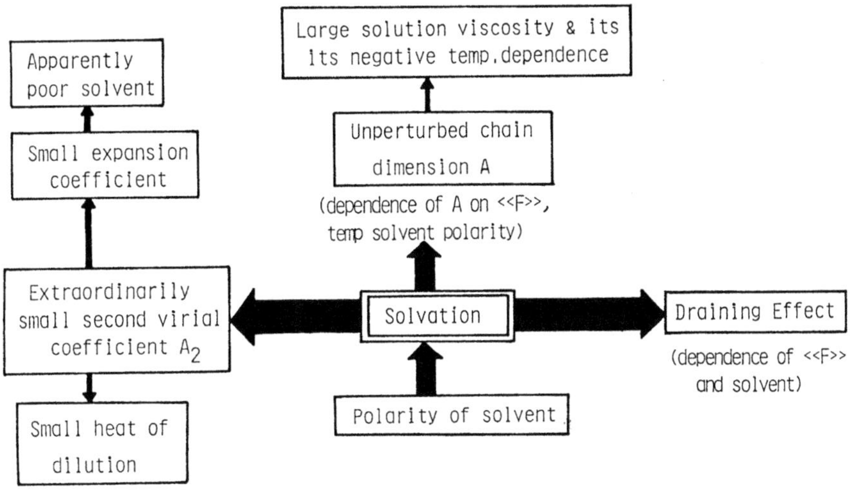

Fig. 42. Correlation between solvation and some characteristic features of solutions of cellulose derivatives

between C_3—OH and O_5' (ring oxygen), and between C_6—OH and bridge oxygen, and (3) the steric hindrance of the solvating solvent molecules. In Fig. 40, the region I is that due to the steric hindrance of the pyranose ring without substituent group only. Fig. 41 shows that the flexibility of the CA chain becomes larger in polar solvents. These experimental results preclude that the inter-molecular hydrogen bonds are more or less preserved in CA solutions and contribute significantly to the rigidity of the chain. Hence the factor (2) can be ignored, at least if CA solutions are concerned. The region II in Fig. 40 represents qualitatively the contribution of the steric hindrance due to the substituent group (O-acetyl group), although the above three factors are not simply additive to give the A value. The region III can be roughly regarded as the contribution of the solvation to A, because the steric effect of the solvents shown in the Figure is of the same order. It can be seen that, if CA is dissolved in a highly polar solvent, the solvation effect plays an important role in determining the rigidity of the CA molecule. In fact lyotropic liquid crystals are formed when CA is dissolved in highly polar solvents or in acids [125].

The characteristic features of dilute solutions of cellulose derivatives can be reasonably and consistently explained by solvation (Fig. 42).

13 Solubility of Cellulose in Aqueous Alkali Solution: Explanation by the Intra-Molecular Hydrogen Bond Concept

Since Mercel discovered the process of the so-called mercerization of cellulose (a process leading to cotton fabrics by treating cellulose under tension with an aqueous alkali solution), the swelling phenomena of cellulose in an aqueous alkali solution have been studied in detail, by many investigators. For example, for natural cellulose, about 8–10 wt% aqueous sodium hydroxide (NaOH) at low temperature was found experimentally to be the most powerfull swelling agent [126], in which only a small part (probably a low molecular weight fraction) of cellulose is soluble [127–130]. Staudinger et al. [129] described, without giving details, that cotton and mercerized cotton can be dissolved in a 10% (w/v) NaOH solution, if the viscosity average degree of polymerization (P_v) is below 400, and regenerated cotton if P_v is less than 1200. The solubility behavior of the regenerated cellulose observed by Staudinger et al. was, unfortunately, not reproducible. Viscose and cuprammonium rayons never dissolve completely in 10% (w/v) aqueous NaOH. The expression "Löslichkeit in 10-iger NaOH", which Staudinger used in his study should be understood to mean partial dissolution. The experimental fact that cellulose contains an alkali soluble part was used to evaluate the lateral order distribution of cellulose fibers, without seriously considering the molecular weight fractionation effect by the alkali. It was accepted without direct evidence, that the difference in the aggregate state of cellulose molecular chains predominantly influences the solubility of cellulose towards alkali. From a theoretical point of view this is only true if there are no intra- and inter-molecular hydrogen bonds. Recent advance in the high resolution NMR technique, especially in the cross-polarization magic angle technique for solid cellulose [131] will facilitate the solution of the problem of the solubility of cellulose in aqueous NaOH.

Kamide et al. [29] were able to prepare cellulose samples having fairly high molecular weights, which dissolved to a large extent in a dilute alkali solution at 4 °C. The solubility was tested as follows [29]: cellulose film samples were scissored into small pieces (5 mm × 5 mm) and vacuum-dried at 40 °C for 18 h. The cellulose (1–5 parts) was dispersed into 1–15 wt% aqueous NaOH (99–95 parts) precooled at 4 °C, for 1 h at 4 °C, using a home mixer intermittently (1 min) to minimize the rise in local temperature. The dispersed solution was then ultracentrifuged at 20,000 rpm, at 4 °C for 45 min, followed by measuring the amount of cellulose (m_G) in the remaining gelatinous layer (lower liquid phase) by regeneration. The solubility of cellulose S_a is defined by Eq. (59):

$$S_a = (m_C - m_G)/m_C \tag{59}$$

where m_C denotes the amount of cellulose originally used in this test.

Table 15. Molecular characteristics of cellulose samples [29]

Sample	Crystal form	$M_v \times 10^{-4}$	S_a[a] (%)	χ_{am}(X)	χ_{ac}(IR)	χ_h(NMR)
BRC-1	II	8.1	52	0.46	0.59	0.56
BRC-2	II	7.9	62	0.69	0.68	0.62
BRC-3	II	7.9	81	0.88	0.64	0.79
BRC-4	II	7.9	90	0.73	0.68	0.87
BRC-5	II	7.9	100	0.94	—	—
CL-1	I	19.4	9	0.21	—	—
CL-2	I	8.3	31	0.25	—	—
CL-3	I	5.8	30	0.26	—	—
PC-1	I	1.9	58	0.92	—	—

[a] Measured at initial polymer concentration c = 5 wt% for a 10 wt% aq. NaOH at 4 °C

A cellulose sample, prepared physically from pulp, dissolved by as much as 58% (at c = 5%) in alkali and a significant increase in amorphous content was observed, although M_v was as low as 1.9×10^4. Accordingly, an increase in the solubility of cellulose having crystal form I seems to reflect an increase in amorphous content or a decrease in M_v. However, no physical treatment can produce cellulose having a solubility S_a (at c = 5 wt%) >0.85. Kamide et al. [29] disclosed the relationship between S_a and the amorphous fraction. Table 15 summarizes the crystal form, M_v, S_a,

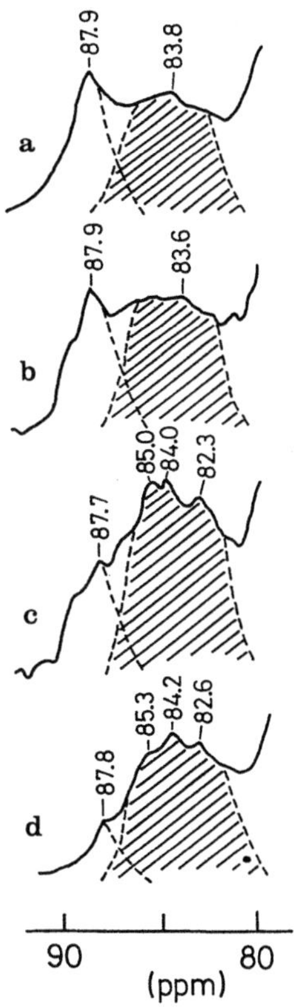

Fig. 43 a–d. CP-MASS ^{13}C-NMR spectra of the C_4 carbon peak region for the BRC series samples; **a** BRC-1; **b** BRC-2; **c** BRC-3; **d** BRC-4. Numbers on the peaks denote peak values in ppm, broken lines denote separation of peak areas under the peaks. Hatched area shows peak area for higher magnetic field components [29]

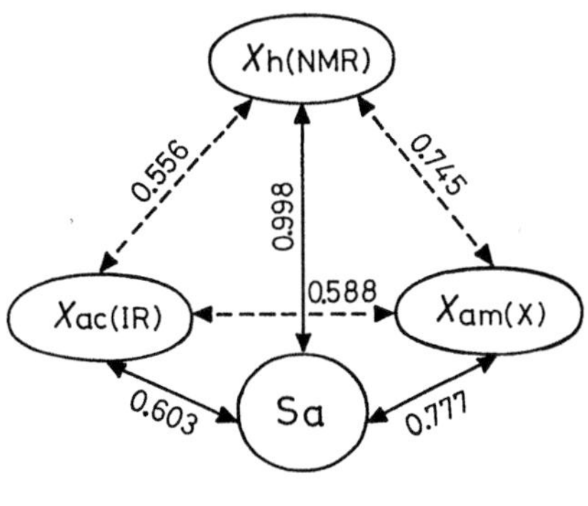

Fig. 44. Correlations among solubility S_a and the so-called amorphous content evaluated by X-ray diffraction, IR and NMR [29]. Numbers on the lines are the correlation coefficients γ for two arbitrarily chosen parameters

and the amorphous fraction of the investigated samples. Sample PC-1 was almost amorphous cellulose, as indicated by X-ray diffraction, but its S_a value was only 0.58. Thus, $\chi_{am}(X)$ ($\equiv 1 - \chi_c(X)$, $\chi_c(X)$ being the crystallinity determined by X-ray diffraction, is not the only factor governing S_a.

Next, a correlation among S_a, $\chi_{am}(X)$, the fraction of the accessible part (by heavy water) at equilibrium, as determined by the deuteration IR method, $\chi_{ac}(IR)$, and the relative amount of the higher field peaks of the C_4 carbon peaks, $\chi_h(NMR)$, were examined for samples having cellulose crystal form II, so as to clarify the factors contributing to S_a. In this case, $\chi_h(NMR)$ was estimated from Eq. (60):

$$\chi_h(NMR) = I_h/(I_l + I_h) \tag{60}$$

where I_l and I_h denote the areas under the peaks centered at 87.9 and 84 ppm, respectively, and the line shapes were assumed to be Lorentzian (Fig. 43).

Figure 44 shows the correlation among S_a, $\chi_{am}(X)$, $\chi_{ac}(IR)$, $\chi_h(NMR)$. The numbers on the lines are the correlation coefficients γ between two arbitrarily chosen parameters; γ for $S_a - \chi_h(NMR)$ is in fact as high as 0.998 and for $S_a - \chi_{am}(X)$ and $S_a - \chi_{ac}(IR)$, γ is 0.777 and 0.603, respectively. Thus, $\chi_h(NMR)$ is most closely related to S_a. In other words, the solubility behavior of cellulose cannot be explained by the concepts of "crystal-amorphous" or "accessible-inaccessible" only. The correlation coefficients γ between the so-called amorphous content parameters are as follows: $\chi_{am}(X) - \chi_h(NMR)$, 0.745; $\chi_h(NMR) - \chi_{ac}(IR)$, 0.556; $\chi_{am}(X) - \chi_{ac}(IR)$, 0.588.

Figure 45 shows the chemical structure of a cellobiose unit. All C_1' and C_4 carbons are linked by oxygen atoms which are the center of rotation accompanied by conformational change in the pyranose ring. Thus, one major factor influencing the magnetic properties of the C_1 and C_4 carbons is conformational change along the C_1'—O—C_4 sequence. This change is predominantly controlled by intramolecular hydrogen bonding between the neighboring glucopyranose units in a cellobiose unit. An intramolecular hydrogen bond between the hydrogen of the OH group attached to the C_3 carbon and the ring oxygen in the neighboring glucopyranose ring causes the formation of a seven membered cyclic ether; the C_3 and C_4 carbons are located at the terminals of a kind of conjugated system of electrons involving the oxygen atoms and the C—O bonds. If the electrons in this system are mobile, the C_4 carbon is cationized and the C_3 carbon is anionized. If the intermolecular interaction (which can be disregarded as neglibible even if it exists) is not taken into account, the electron density on the C_4 carbon becomes lower than that on a C_4 carbon not participating in seven-membered ether formation, resulting in a shift of the C_4 carbon NMR peak to a lower magnetic field. Hence, we can consider a partial breakdown of the intramolecular hydrogen bonds existing in the cellulose sample to create a magnetic field around the C_4 carbon

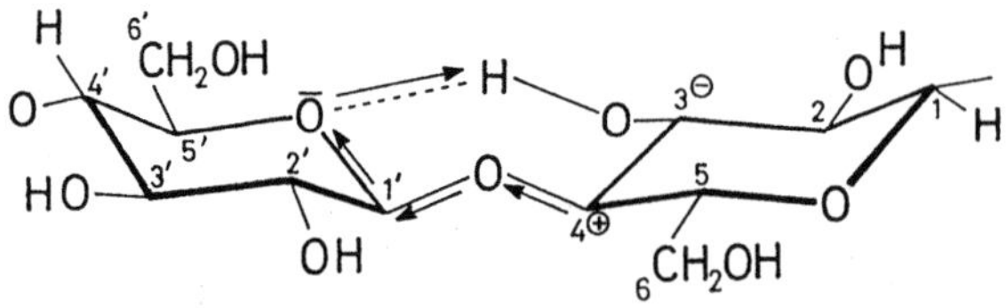

Fig. 45. Schematic representation of the π-σ electron conjugate system in a cellobiose unit [29]

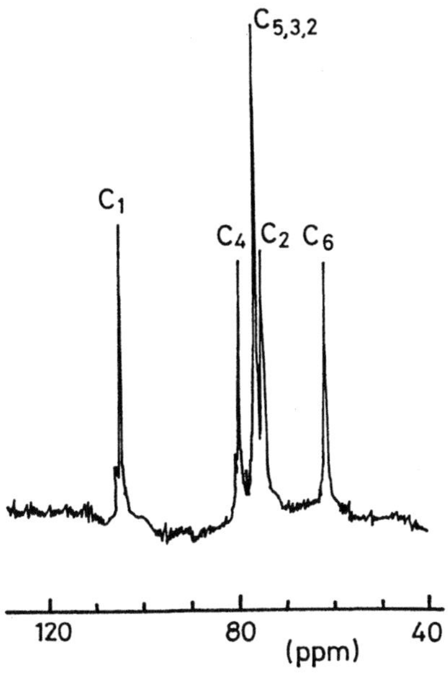

Fig. 46. ^{13}C-NMR spectrum for BRC-4 dissolved in NaOH—D_2O (1:9, w/w) [29)]

heterogeneously, so that its peak (and possibly also that of the C_1 carbon) produces a new component at higher magnetic field, compensating for the sharp component of the original C_4 carbon peak. This change results in the formation of two envelopes in the C_4 carbon peak region. The systematic change observed in Fig. 43 can be reasonably interpreted by the above mechanism.

It may be concluded that the low field envelope in the C_4 carbon peak can be assigned to cellobiose units having fewer intramolecular hydrogen bonds (i.e., the region around such cellobiose units in cellulose molecules has a very disordered conformation). We cannot simply conclude that in the absence of any intramolecular hydrogen bonds, the peak would narrow significantly, since cellobiose units in cellulose molecules in the solid state can take on a large number of conformations.

On the basis of the above mechanism, a variation in the lowest peak intensity among the C_1 carbon peaks between cellulose samples can be explained in the same manner as that of the C_4 carbon peak, i.e., the breakdown of intramolecular hydrogen bonds. The lowest peak (107.3 ppm) for the C_1 carbon becomes weak in the following order: BRC-1 > BRC-2 > BRC-3 > BRC-4.

Figure 46 shows the ^{13}C-NMR spectrum for the solution of BRC-4 in NaOH—D_2O (1:9, w/w). Except for the C_3 and C_5 carbon peaks, all others shift to a magnetic field higher than that for solid BRC-4. One C_4 carbon peak for the solution is located at 79.9 ppm, which is a much higher magnetic field than that (87.9 and about 84 ppm) for the original BRC-4 solid. BRC-4 dissolves molecularly in 10 wt % aqueous NaOH, without forming an alcoholate, and solvating with solvent. Hence, no intramolecular hydrogen bonds should exist in the solution. The appearance of the C_4 carbon peak at 79.9 ppm as a single sharp peak in the solution is quite reasonable considering that the C_4 carbon peak must shift to a higher magnetic field when the intramolecular hydrogen bonds are destroyed.

In view of the chemical structure of the carbon-oxygen sequences around the C_4 and C_5 carbons, the deshielding effects on the C_4 and C_5 carbon may be expected to be of the same order when the intramolecular hydrogen bonds are completely destroyed. Thus, the C_4 carbon peak should appear near the C_5 carbon peak. The C_4 carbon peak for the BRC-4 solution was observed at 79.95 ppm, and this is very close to the C_5 carbon peak at 76.5 ppm. Destruction of the intramolecular hydrogen bonds is expected to cause the C_3 carbon peak for solid cellulose to shift to a lower magnetic field, which is the reverse of the shift of the C_4 carbon peak. The main C_3 carbon peak for the original ERC-1 is located at 75.1 ppm, while the C_3 carbon peak for the BRC-4 solution in alkali is oberved at 76.4 ppm. The latter corresponds to a slightly lower field than the former.

Kamide et al. attempted to analyze the CP/MASS ^{13}C-NMR peaks of each carbon of various samples having crystalline forms of cellulose I or II, with regard to intra- and intermolecular hydrogen bonds [132]. The spectrum of ball milled cellulose powder had broad carbon peaks at almost the same positions as those in solution, suggesting that there are no strong intra-molecular hydrogen bonds with a specific bond length. For cellulose I and II, one or two sharp components, respectively, in addition to a broad component, were observed for each carbon. The former components were assigned to strong intra-molecular hydrogen bonding. There are three possible intra-molecular hydrogen bonds, O_3—H ... O_5', O_2—H ... O_6', and O_6—H ... O_2'; simultaneous occurence of the first two hydrogen bonds can explain consistently all peaks experimentally observed in the NMR spectrum of cellulose I. To explain the NMR spectrum of cellulose II all three intra-molecular hydrogen bonds are nessesary.

Using the same technique of CP/MASS ^{13}C-NMR, Kamide et al. [133] disclosed that in aged solid alkali cellulose the molecular motion is relatively rapid, as compared with solid cellulose; intramolecular hydrogen bonding is present, and the sodium ion selectively coordinates to the hydroxyl oxygen at the C_2 position of cellulose.

14 Concluding Remarks

1. There are close relationships between the distribution of the degree of substitution ($\langle\langle f_k \rangle\rangle$, $\langle\langle F \rangle\rangle$, and $g(\langle F \rangle)$) and the physical properties of cellulose derivatives (CD); we are presently approaching the goal to design the particular molecular structure of CD required for a desired physical property (for example, anti-coagulant activity or solvent absorbing power).
2. Almost molecularly monodisperse CD samples can be prepared using the successive solution fractionation method.
3. CD molecules dissolve in solvent, with solvation. Usually, solution is better achieved lowering the temperature.
4. There exists an upper and lower critical solution temperature for same CD solutions.
5. CD molecules are partially free draining chains and the negative temperature dependence of the limiting viscosity number $[\eta]$ can be attributed to the temperature dependence of the unperturbed chain dimension A.
6. The exponent a is the Mark-Houwink-Sakurada equation ($[\eta] = K_m M_w^a$) is mostly smaller than 0.8.

7. The unperturbed chain dimension A is relatively large, being larger in solvents having higher dielectric constants ε: The flexibility of the CD molecule depends on the polarity of the solvent.
8. Cellulose acetate (CA) molecules become most rigid at $\langle\langle F \rangle\rangle \simeq 2.5$, if they are dissolved in a common solvent (DMAc or DMSO). By doubly extrapolating A values of CA to $\langle\langle F \rangle\rangle = 0$ and $\varepsilon = 1$, the cellulose molecule, hypothetically dissolved in a non-polar solvent, should be expected to be a freely rotating chain.
9. CD are generally semi-flexible chains.
10. The "Two-Parameter Theory" cannot explain the dilute solution properties of CD, because in CD solutions solvatation and the draining effect can never be neglected.
11. The solvents suitable for CD are apparently poor (small A_2 and $\alpha_s \simeq 1$), but the CD solutions are stabilized by solvatation, and the CD molecules are expanded.

15 References

1. Staudinger, H., Freudenberger, H.: Ber. dtsch. chem. Ges. *63*, 2331 (1930)
2. Kamide, K., Miyazaki, Y.: Polym. J. *10*, 409 (1978)
3. Kamide, K., Okajima, K.: Polym. J. *13*, 127 (1981)
4. Gardner, T. S., Purves, C. B.: J. Am. Chem. Soc. *64*, 1539 (1942)
5. Malm, C. J., Tanghe, J. J., Laird, B. C.: J. Am. Chem. Soc. *72*, 2674 (1950)
6. Goodlett, V. W., Dougherty, J. T., Patton, H. W.: J. Polym. Sci., A-1 *9*, 155 (1971)
7. Kamide, K., Saito, M.: Eur. Polym. J. *20*, 903 (1984)
8. Wu, T. K.: Macromol. *13*, 74 (1980)
9. Kamide, K., Okajima, K.: Polym. J. *13*, 163 (1981)
10. Kamide, K., Okajima, K., Kowsaka, K., Matsui, T., Nomura, S., Hikichi, K.: Polym. J. *17*, 909 (1985)
11. Kamide, K., Manabe, S., Osafune, E.: Makromol. Chem. *168*, 173 (1973)
12. Kamide, K., Matsui, T., Okajima, K., Manabe, S.: Cell. Chem. Technol. *16*, 601 (1982)
13. Kamide, K., Okada, T., Terakawa, T., Kaneko, K.: Polym. J. *10*, 547 (1978)
14. Bergström, S.: Naturwissenschaften *25*, 706 (1935)
15. Astrup, T., Galsmar, I., Volkert, M.: Acta Physiol. Scand. *8*, 215 (1944)
16. Karrer, P.. Köenig, H., Usteri, E.: Helv. Chim. Acta *26*, 1296 (1943)
17. Astrup, J., Piper, J.: Acta Physiol. Scand. *9*, 351 (1945)
18. Piper, J.: Acta Physiol. Scand. *9*, 28 (1945)
19. Piper, J.: "Farmakologiske Undersögelser över Syntetiske Heparin-lignande Stoffer", (Disp.) Copenhagen, 1945
20. Felling, J., Wiley, C. E.: Arch. Biochem. Biophys. *85*, 313 (1959)
21. Rothschild, A. M.: J. Pharmac. Chemother. *33*, 501 (1968)
22. Rothschild, A. M., Castania, A.: J. Pharm. Pharmac. *20*, 77 (1968)
23. Kiss, J.: "Chemical Structure of Heparin", in "Heparin", Thomas, K. ed., Academic Press, London, 1976, p9
24. Kamide, K., Okajima, K., Matsui, T., Ohnishi, M., Kobayashi, H.: Polym. J. *15*, 309 (1983)
25. Kamide, K., Okajima, K., Matsui, T., Kobayashi, H.: Polym. J. *16*, 259 (1984)
26. Jansen, E.: Ger. Pat., 332, 203 (1918)
27. Callihan, C. D.: "Cellulose Technology Research", Turbak, A. F. ed., ACS series 10, 1975, p33
28. Ott, E., Spurlin, H. M.: "Cellulose and Cellulose Derivatives", vol. II, 1954, John Wiley & Sons Inc., p944
29. Kamide, K., Okajima, K., Matsui, T., Kowsaka, K.: Polym. J. *16*, 857 (1984)
30. See, for example, Cragg, L. H., Hammerschlag, H.: Chem. Rew. *39*, 79 (1946), Hall, R. W., "Techniques of Polymer Characterization", Allen, P. W. ed., 1958, Butterworth London, Chp. 2, Guzman, G. M., "Progress in High Polymers", Robb, J. C., Peader, F. W. ed., 1961,

Heywood London, Vol. 1, p113, Cantow, M. J. R. ed., "Polymer Fractionation", 1967, New York, Academic Press, Kawai, T. ed., "Polymer Engineering", 1967, Vol. 4, Chijin Shokan, Tokyo, Tung, L. H. ed., "Fractionation of Synthetic Polymers", 1977, Marcel Dekker, New York
31. Kamide, K.: "Fractionation of Synthetic Polymers", Tung, L. H. ed., 1977, Marcel Dekker, New York, Chp. 2
32. Kamide, K., Miyazaki, Y., Abe, T.: Makromol. Chem. *117*, 485 (1976)
33. Kamide, K., Miyazaki, Y., Abe, T.: Polym. J. *9*, 395 (1977)
34. Kamide, K., Matsuda, S., Miyazaki, Y.: Polym. J. *16*, 479 (1984)
35. Kamide, K., Matsuda, S.: Polym. J. *16*, 515 (1984)
36. Kamide, K., Matsuda, S.: Polym. J. *16*, 591 (1984)
37. Kamide, K., Miyazaki, Y.: Polym. J. *12*, 153 (1980)
38. Kamide, K., Miyazaki, Y., Abe, T.: Brit. Polym. J. *13*, 168 (1981)
39. Kamide, K., Miyazaki, Y., Abe, T.: Polym. J. *11*, 523 (1979)
40. Kamide, K., Terakawa, T., Miyazaki, Y.: Polym. J. *11*, 285 (1979)
41. Saito, M.: Polym. J. *15*, 249 (1983)
42. Kamide, K., Saito, M., Abe, T.: Polym. J. *13*, 421 (1981)
43. See, for example, Nair, P. R. M., Gohil, R. M., Patel, K. C., Patel, R. D.: Eur. Polym. J. *13*, 273 (1977), Lachs, H., Kronman, K., Wajs, J.: Kollid Z. *79*, 91 (1937), Levi, G. R., Giera, A.: Gazz, Chim. Ital. *67*, 719 (1937), Levi, G. R., Villota, U., Montirelli, M.: Gazz. Chim. Ital. *68*, 589 (1938), Bezzi, S., Croatta, U.: Atti Inst. Veneto Sci. *99*, 905 (1939–1940), Munster, A.: J. Polym. Sci. *5*, 333 (1950), Sobue, H., Matsuzaki, K., Yamakawa, K.: Sen'i Gakkaishi *12*, 100 (1956)
44. See, for example, Hunt, M. L., Newman, S., Scheraga, H. A., Flory, P. J.: J. phys. Chem. *60*, 1278 (1956), Manley, R. S. J.: Ark. Kemi. *9*, 519 (1956), Huque, M. M., Goring, D. A., Mason, S. G.: Can. J. Chem. *36*, 952 (1958), Tanner, D. W., Berry, G. C.: J. Polym. Sci. C *12*, 941 (1974), Holtzer, A. M., Benoit, H., Doty, P.: J. phys. Chem. *58*, 624 (1954)
45. Kamide, K., Saito, M.: Polym. J. *14*, 517 (1982)
46. See, for example, Wales, M., Swanson, D. L.: J. Phys. & Colloid Chem. *55*, 203 (1951), Sperling, L. H., Easterwood, M.: J. Appl. Polym. Sci. *4*, 25 (1960), Bikales, N. M., Segal, L. ed., "High Polymers", Vol. V, Part IV, John Wiley & Sons, Inc., New York, 1971, p433, Klenin, V. J., Denisova, G. P.: J. Polym. Sci., Polym. Symp. No. *42*, 1563 (1973), Goeble, K. D., Berry, G. C.: J. Polym. Sci., Polym. Phys. Ed. *15*, 555 (1977)
47. Kamide, K., Terakawa, T., Manabe, S.: Sen'i Gakkaishi *30*, T-464 (1974)
48. Kamide, K., Terakawa, T., Manabe, S., Miyazaki, Y.: Sen'i Gakkaishi, *31*, T-410 (1975)
49. Kamide, K., Manabe, S., Terakawa, T.: JP885, 873 (1977), and 909, 158 (1978)
50. Kamide, K., Terakawa, T., Uchiki, H.: Makromol. Chem. *177*, 1447 (1976)
51. Kamide, K., Terakawa, T., Matsuda, S.: Brit. Polym. J. *15*, 91 (1983)
52. Ikeda, T., Kawaguchi, H.: Rep. Prog. Polym. Sci., Jpn. *9*, 23 (1966)
53. Cowie, J. M., Ranson, R. J.: Makromol. Chem. *143*, 105 (1971)
54. Kamide, K., Abe, T., Miyazaki, Y., Watanabe, M.: J. Soc. Text. Masch. Jpn. *34*, T-1 (1981)
55. Suzuki, H., Ohno, K., Kamide, K., Miyazaki, Y.: Netsusokutei (Calor. therm. Analysis) *8*, 67 (1981)
56. Suzuki, H., Kamide, K., Saito, M.: Eur. Polym. J. *18*, 123 (1980)
57. Suzuki, H., Muraoka, Y., Saito, M., Kamide, K.: Brit. Polym. J. *14*, 23 (1982)
58. Kamide, K., Manabe, S.: "Material Science of Synthetic Membranes; Role of Microphase Separation Phenomena in the Formation of Porous Polymeric Membrane", Loyed, D. R. ed., ACS Symposium Series 269, ACS, Washington D.C., 1985, Chp. 9, p197
59. Kamide, K., Matsuda, S.: Polym. J. *16*, 825 (1984)
60. Kamide, K., Matsuda, S., Saito, M.: Polym. J. *17*, 1013 (1985)
61. Kamide, K., Matsuda, S.: unpublished results
62. Koningsveld, R., Kleintjens, L. A., Shultz, A. R.: J. Polym. Sci., A-2 *8*, 126 (1970)
63. Shultz, A. R., Flory, P. J.: J. Am. Chem. Soc. *74*, 4760 (1952)
64. See, for example, Kurata, M.: "Industrial Chemistry of High Polymers", Vol. III, Modern Industrial Chemistry Ser. No. 18, Asakura, Tokyo, 1975, Chp. 4
65. Kurata, M., Fukatsu, M., Sotobayashi, H., Yamakawa, H.: J. Chem. Phys. *41*, 139 (1964)
66. Fixman, M.: J. Chem. Phys. *36*, 3123 (1962)
67. Saito, M.: Polym. J. *15*, 213 (1983)

68. Kamide, K., Saito, M.: Eur. Polym. J. *19*, 507 (1983)
69. Kamide, K., Saito, M., Suzuki, H.: Makromol. Chem., Rapid Commun. *4*, 33 (1983)
70. Spurlin, H. M.: "Cellulose and Cellulose Derivatives", Otta, E. ed., Interscience Pub., New York, N.Y., 1943, p868
71. Clermont, P.: Ann. Chim. *12*, 2420 (1943)
72. Marsden, R. J. B., Urquhart, A. R.: J. Text. Inst. *33*, T-105 (1942)
73. Kamide, K., Okajima, K., Saito, M.: Polym. J. *13*, 115 (1981)
74. See, for example, Passynsky, A.: Acta physicochem., USSR *22*, 137 (1947), Passynsky, A.: J. Polym. Sci. *29*, 61 (1958)
75. Moore, W. R., Tidswell, B. M.: Makromol. Chem. *81*, 1 (1965)
76. Moore, W. R.: J. Polym. Sci., Part C, No16, 571 (1967)
77. Moore, W. R.: "Solution Properties of Natural Polymers, An International Symposium", The Chemical Society Burlington House, London, W.1, 1968, Group 3, p185
78. Schulz, G. V., Penzel, E.: Makromol. Chem. *112*, 260 (1968)
79. Penzel, E., Schulz, G. V.: Makromol. Chem. *113*, 64 (1968)
80. Shanbhag, V. P.: Ark, Kemi *29*, 1 (1968)
81. Shanbhag, V. P.: Ark. Kemi *29*, 139 (1968)
82. Brown, W., Henley, D., Ohman, J.: Makromol. Chem. *64*, 49 (1963)
83. Manley, R. S.: Ark, Kemi *9*, 519 (1956)
84. Henley, D.: Ark. Kemi *18*, 327 (1961)
85. Valtasaari, L.: Makromol. Chem. *150*, 117 (1971)
86. Krigbaum, W. R., Sperling, L. H.: J. phys. Chem. *64*, 99 (1960)
87. Neely, W. B.: J. Polym. Sci., A *1*, 311 (1963)
88. Das, B., Ray, A. K., Choudhurry, P. K.: J. phys. Chem. *73*, 3413 (1969)
89. Das, B., Choudhury, P. K.: J. Polym. Sci., A-1 *5*, 769 (1967)
90. Brown, W., Henley, D.: Makromol. Chem. *79*, 68 (1964)
91. Kurata, M., Tsunashima, Y., Iwata, M., Kamata, K.: "Polymer Handbood", 2nd ed., Brandrup, J., Immergut, E. H. ed., John Wiley & Sons, New York, N.Y., 1975
92. Kamide, K., Miyazaki, Y., Abe, T.: Makromol. Chem. *180*, 2801 (1979)
93. Nair, P. R. M., Gohil, R. M., Patel, K. C., Patel, R. D.: Eur. Polym. J. *13*, 273 (1977)
94. Dymarchuk, N. P., Mishchenko, K. P., Fomia, T. V.: Zhur. Prikl. Khim. (Leningrad), *37*, 2263 (1964)
95. Shakhparonov, M. I., Zahurdayeva, N. P., Podgarodetshii, Ye. K.: Vysokomol. Soedin., Ser. A, *9*, 1212 (1967)
96. Staudinger, H., Eicher, T.: Makromol. Chem. *10*, 261 (1953)
97. Sharples, A., Major, H. M.: J. Polym. Sci. *27*, 433 (1958)
98. Howard, P., Parikh, S. S.: J. Polym. Sci., Part A-1 *4*, 407 (1966)
99. Ishida, S., Komatsu, H., Kato, H., Saito, M., Miyazaki, Y., Kamide, K.: Makromol. Chem. *183*, 3075 (1982)
100. Singer, S. J.: J. Chem. Phys. *15*, 341 (1947)
101. Holmes, F. H., Smith, D. I.: Trans. Faraday Soc. *53*, 69 (1957)
102. Golubev, V. M., Frenkel, S. Y.: Vysokomol. Soedin., Ser. A, *10*, 750 (1968)
103. Howlett, F., Minshall, E., Urquhart, A. R.: J. Text. Inst. *35*, T133 (1944)
104. Mandelkern, L., Flory, P. J.: J. Am. chem. Soc. *74*, 2517 (1952)
105. Flory, P. J., Spurr, O. K. Jr., Carpenter, D. K.: J. Polym. Sci. *27*, 231 (1958)
106. Moore, W. R., Brown, A. M.: J. Colloid Sci. *14*, 1 (1959)
107. Moore, W. R., Brown, A. M.: J. Colloid Sci. *14*, 343 (1959)
108. Moore, W. R., Edge, G. D.: J. Polym. Sci. *47*, 469 (1960)
109. Kamide, K., Ohno, K., Kawai, T.: Kobunshi Kagaku *20*, 151 (1963)
110. Suzuki, H., Miyazaki, Y., Kamide, K.: Eur. Polym. J. *16*, 703 (1980)
111. See, for example, Yamakawa, H.: "Modern Theory of Polymer Solutions", Harper & Row, New York, 1971, Chp. III, p91
112. Kurata, M., Stockmayer, W. H.: Fortschr. Hochpolym. Forsch. *3*, 196 (1963)
113. Kurata, M., Yamakawa, H., Teramoto, E.: J. Chem. Phys. *28*, 785 (1958)
114. Baumann, H.: J. Polym. Sci., Polym. lett. *3*, 1069 (1965)
115. Stockmayer, W. H., Fixman, M.: J. Polym. Sci., Part C, *1*, 137 (1963)
116. Kamide, K., Moore, W. R.: J. Polym. Sci., Polym. lett. *2*, 1029 (1964)

117. Cowie, J. M., Bywater, S.: Polymer *6*, 197 (1965)
118. Kamide, K., Miyazaki, Y.: Polym. J. *10*, 539 (1978)
119. Kamide, K., Saito, M.: Eur. Polym. J. *17*, 1049 (1981)
120. Kamide, K., Saito, M.: Eur. Polym. J. *18*, 661 (1982)
121. Benoit, H., Doty, P. M.: J. Phys. Chem. *57*, 958 (1953)
122. Yamakawa, H., Stockmayer, W. H.: J. Chem. Phys. *57*, 2843 (1972)
123. Kirkwood, J., Riseman, J.: J. Chem. Phys. *16*, 565 (1948)
124. Yamakawa, H., Fujii, M.: Macromol. *7*, 128 (1974)
125. Kamide, K., Okajima, K., Matsui, T.: USP 4,370,168 (1980)
126. See, for example, Neals, S. M.: J. Text. Inst. *20*, T373 (1920), Beadle, C., Stevens, H. P., the 8th International Congress on Applied Chemistry, Vol. 13, 1912, p25
127. Dillenius, H.: Kunstseide Zellwolle *22*, 314 (1940)
128. Schwart, E., Zimmerman, W.: Melliads Textileber. Int. *22*, 525 (1941)
129. Staudinger, H., Mohr, R.: J. Prakt. Chem. *158*, 233 (1941)
130. Eisenluth, O.: Cellulose Chem. *19*, 45 (1941)
131. See, for example, Atalla, R. H.: J. Am. chem. Soc. *102*, 3249 (1980), Earl, W. L., Vander Hart, D. L.: J. Am. chem. Soc. *102*, 3251 (1980), Earl, W. L., Vander Hart, D. L.: Macromol. *14*, 570 (1981), Hirai, F., Horii, A., Kitamaru, R.: Polym. Prepr. Jpn. *31*, 842 (1982), Hirai, F., Horii, A., Akita, A., Kitamaru, R.: Polym. Prepr. Jpn. *31*, 2519 (1982), Hirai, F., Horii, A., Kitamaru, R.: Prepr. 47th Chem. Soc. Jpn. Ann. Meeting, 1983, p1392, Hayashi, J., Takai, M., Tanaka, R., Hatano, M., Mozawa, S.: Prepr. 47th Chem. Soc. Jpn. Ann. Meeting, 1983, p1393, Maciel, G. E.: Macromol. *15*, 686 (1982)
132. Kamide, K., Kowsaka, K., Okajima, K.: Polym. J. *17*, 701 (1985)
133. Kamide, K., Okajima, K., Kowsaka, K., Matsui, T.: Polym. J. *17*, 707 (1985)

Editors: G. Henrici-Olivé and S. Olivé
Received March 24, 1986

Structural and Mechanical Properties of Biopolymer Gels

Allan H. Clark and Simon B. Ross-Murphy
Unilever Research, Colworth House, Sharnbrook, Bedford MK44 1LQ U.K.

The structural and mechanical properties of gels formed from biopolymers are discussed both in terms of the techniques used to characterise these systems, and in terms of the systems themselves. The techniques included are spectroscopic, chiroptical and scattering methods, optical and electron microscopy, thermodynamic and kinetic methods and rheological characterisation. The systems considered are presented in order of increasing complexity of secondary, tertiary and quaternary structure, starting with gels which arise from essentially 'disordered' biopolymers via formation of 'quasicrystalline' junction zones (e.g. gelatin, carrageenans, agarose, alginates etc.), and extending to networks derived from globular and rod-like species (fibrin, globular proteins, caseins, myosin) by a variety of crosslinking mechanisms. Throughout the text, efforts are made to pursue the link (both from experiment and from theory) between the structural methods and mechanical measurements. As far as we are aware this is the first major Review of this area since that of J. D. Ferry in 1948 – The interest shown by polymer physicists in more complex biochemical systems, and the multi-disciplinary approaches now being applied in this area, make the format adopted here, in our opinion, the most logical and appropriate.

Advances in Polymer Science 83

Glossary of Symbols and Abbreviations

BST	Blatz, Sharda and Tschoegl (equation)
CD	circular dichroism
CSK	Coniglio, Stanley and Klein (theory)
DCF	density correlation function
DLVO	Deryjaguin-Landau-Verwey-Overbeck (theory)
d.s.c.	differential scanning calorimetry
FRS	forced Rayleigh scattering
FS	Flory-Stockmayer (theory)
M/G	mannuronic/galacturonic acid ratio (alginates)
NMR	nuclear magnetic resonance
ORD	optical rotatory dispersion
i.r.	infrared spectroscopy
QELS	quasi-elastic light scattering
UV	ultraviolet spectroscopy

a	generalised front factor
C	concentration
C_0	critical gelling concentration
C*	chain overlap concentration
C^0	initial concentration
C_1, C_2	Mooney-Rivlin coefficients
D	Hausdorff dimension
D	Z-average translational diffusion coefficient
D_{app}	apparent D measured at finite q
D_m	mutual diffusion coefficient
D_{self}	self diffusion coefficient
D_Z	effective diffusion coefficient
DP_n	number average degree of polymerisation
E	Young's modulus
f	friction coefficient
f	functionality
f_w	weight-average functionality
f^1	fraction of residues in ordered conformation
g	rubber elasticity front factor
G	shear modulus
G′	real part of oscillatory shear modulus
G″	imaginary part of oscillatory shear modulus
G*	complex oscillatory shear modulus
G_e	equilibrium shear modulus
h	fraction of residues in helical conformation
h_g	value of h at the gel point
ΔH_u	enthalpy of gel melting (Takahishi theory)
ΔH^0	van't Hoff enthalpy
J(t)	creep compliance at time t
J_e	equilibrium creep compliance

k	Boltzmann's constant
k_s	Smoluchowski rate constant
M	molecular weight between crosslinks
M_e	entanglement molecular weight
M_j	molecular weight of junction zone
M_{os}	osmotic modulus
M_r	relative molecular mass (molecular weight)
M_w	weight average molecular weight
M_0	initial molecular weight (coagulation kinetics)
n	parameter in BST equation
n	molecularity of junction zones
N_e	number of elastically active junction zones
p	axial (length/diameter) ratio for rods
p_B	bond probability (CSK theory)
q	scattering vector $= (4\pi/\lambda) \sin(\theta/2)$
r	molecular weight ratio (Eq. 36)
R	gas constant
s	order parameter in helix-coil transition theory
$\langle S^2 \rangle_Z$	Z-average mean square radius of gyration
t	time
t_g	gel time
T	absolute temperature
T_m	helix-coil midpoint temperature
T_{mk}	gel melting temperature (°K)
T_2	transverse relaxation time (NMR)
T_τ	effective lifetime of temporary crosslinks
v	extinction probability
V_{max}	maximum rate of enzyme cleavage
V_{mol}	volume per mole of primary chains
$\bar{V}_2$	polymer partial specific volume
W_g	gel fraction
x	number of junction zones per primary chain
α	proportion of reacted functionalities
α_c	critical value of α at the gel point
$[\alpha]$	specific optical rotation
$[\alpha]_t$	value of $[\alpha]$ at time t
γ	shear strain
γ_{max}	maximum shear strain
$\dot{\gamma}$	shear strain rate
δ	loss angle ($\tan \delta = G''/G'$)
δ	exponent in concentration dependence of modulus
ε	tensile strain
ε_B	tensile strain at failure
ζ	correlation length
η	viscosity
$[\eta]$	intrinsic viscosity

θ	scattering angle
λ	wavelength
λ	extension ratio
μ'	constraint release parameter
ξ	statistical weight (CSK theory)
π	osmotic pressure
$\varrho(r)$	density of particles in volume element r from 0
σ	shear stress
σ	helix interruption constant
σ_t	shear stress at time t
σ_T	tensile stress
σ_B	tensile stress at failure
τ	relaxation time
τ	scaled time in coagulation equation
φ	proportionality constant in theory of elasticity
φ	volume fraction
φ_g	critical volume fraction for gelation
φ_p	volume fraction of rods in Flory theory
φ_p^*	critical value of φ_p
φ_s	volume fraction of solvent
ω	oscillatory frequency

1 Introduction

Whilst the properties of synthetic networks and gels are, and have been, the subject of a great deal of work by a number of groups worldwide [1,2], the number of workers in the closely related area of biological gels and networks [3] has until recently been quite few. The same is true even if we extend this class to include systems such as gels formed from extracted and/or modified biopolymers. In general these materials do not involve the formation of extensive covalent crosslinks. Thus in the present Review we shall not consider materials analogous to vulcanized or uncrosslinked natural rubber, (a topic extensively covered by other workers). Rather we shall concentrate upon gels formed from a range of naturally occurring biopolymers, which involve various types of physical crosslink; we also restrict ourselves generally to aqueous solutions, since the aqueous environment is normally essential for the existence and role of biopolymers *in vivo*.

Historically, the major research effort in this whole area has been concentrated upon studying the physico-chemical nature of the crosslinking mechanism, and employing techniques which are more familiar to the biochemist than to the polymer physicist. At the same time the complications introduced by forming gels involving physical rather than covalent crosslinks, should be apparent. It is, in general, much more difficult to establish unambiguously how many crosslinks are formed, and these crosslinks themselves may be, at least to some extent, transient in nature. In this respect biopolymer networks have a great deal in common with systems such as polyethylene in xylene or polyacrylonitrile in dimethylformamide [4], or temporary networks of ionomer systems [5] where interchain bonding is non-covalent. The link is particularly strong with those gels which arise through formation of interchain microcrystallites, particularly the interesting isotactic polystyrene gels extensively studied by Keller, Atkins and co-workers [6,7]. Parallels may also be drawn with the structures formed by other specifically designed synthetic polymer systems. Examples include multiblock macromolecules in solvents of very different quality for the different blocks, and the ionomeric networks mentioned above [5]. A Review by Tsuchida and Abe [8] gives an excellent overview of the disparate mechanisms by which physical crosslinking of synthetic polymer chains can occur.

All of these systems and also the biopolymer networks to be discussed here may be seen at least, superficially, to form a class of materials intermediate between the classical crosslinked elastomer in which a bulk polymer is vulcanized either chemically, thermally or for example using γ-irradiation, and the entanglement network systems which occur either for linear polymers (molecular weight M_r) in the bulk, when the "critical entanglement molecular weight" M_e is exceeded, or for polymer solutions, with $M_r > M_e$, when the concentration of polymer exceeds some critical concentration, C^* [9,10]. Whilst the permanently crosslinked networks have, in general at least one nominally infinite relaxation time (corresponding to an equilibrium modulus G_e), the entanglement networks behave as networks only a short times (when they may be characterised by their plateau moduli). At longer times they relax and show the flow characteristics of a liquid [11].

Another major distinction between most synthetic networks and biopolymer networks is the that latter almost invariably only occur in the presence of excess solvent, i.e. they occur only as solvated networks and, as implied before, this solvent is almost invariably water, or, aqueous electrolyte. This further confuses and compounds the issues, because the thermodynamics of aqueous polymer solutions is itself a very difficult area, the role of solvation forces, the balance of hydrophilicity to hydrophobicity, and the presence of coulombic interactions all contributing to these complications.

A distinction which can naturally be made amongst the various biopolymer gels themselves is that between "*ex vivo*" and "*in vivo*" biogel systems. Whilst, by definition, biopolymers occur in nature, there is also a clear distinction between biopolymer gels which occur themselves, in nature, and those which only form under conditions which are highly antagonistic to most life forms. For example fibrin networks are formed almost spontaneously (except for those suffering from haemophilia) when we cut ourselves, and the formation of the *in vivo* network is the role of the fibrinogen precursor in nature. Similarly the role of algal polysaccharides *in vivo* is to provide shape retention by structuring the interstitial fluid regions of marine plant organisms. This contrasts completely with systems which involve extensive "denaturation" of the biopolymer — for example egg gels formed by heating the protein components (principally ovalbumin) of eggs to above the denaturation temperature (say ~ 80 °C). By definition, for the ovalbumin to retain its native properties it must never be subject to conditions which "denature" it. By contrast, although gelatin is formed by denaturing the triple helical form of the polypeptide tropocollagen, a constituent of connective tissue — intriguingly, gelatin forms gels by partially renaturing to form (in a simple picture) a series of triple helical "junction zones", rather than "point crosslinks". Each of these is separated along the contour of a single chain by regions of flexible polypeptide.

In the above paragraph, we have informally introduced the two major classes of biopolymer which can, with suitable treatment, form gels, viz proteins and polysaccharides. Although in this Review these will be the principal topics for discussion, biopolymers which contain elements of both classes such as glycoprotein systems will also be included. The article will cover both techniques for studying biopolymer gels, and also the systems themselves, and from the above comments it should be clear that we will concentrate upon the many different protein and polysaccharide gel-forming systems and their mechanisms of gelation, as far as is known. Although gels formed from two (or more) different biopolymers, e.g. mixtures of proteins and polysaccharides are currently of great interest, for reasons of space these will not be considered here in any detail. This last area is one which shares a number of features in common with recent studies of synthetic composites and interpenetrating networks [12], but in the biopolymer case the complications introduced by the nature of the components and the presence of water/electrolyte, make this a more complex field, and a sufficiently large topic to require a separate review in order to give justice to the recent literature.

As far as the techniques we shall cover are concerned, these may be divided into (a) structural techniques, including spectroscopy and scattering, microscopy and thermodynamic/thermal methods (which are generally concerned with the mechanism of gelation), and (b) those techniques which are essentially rheological in nature, and are concerned both with small-deformation studies of gelled and gelling systems, and also with their large-deformation and failure properties. When discussing the application of rheological techniques, one of the points which will be considered at some length (in part because of the probable readership of this Review), is the quality of the description of the gel network properties in terms of current models e.g. of rubber elasticity. Whilst the development of treatments for physically cross-linked gels of stiff chains from currently accepted theories for chemically cross-linked gels of flexible chains, and the application of this approach, is an area of only recent interest, some description of the current understanding will be given, since, hopefully the stimulus of work in this area (including that described herein), will catalyse greater interest by both theoreticians and experimentalists.

As far as the organisation of this article is concerned systems will be discussed in the order: gelatin gels, polysaccharide gels, "weak gels" and entanglement networks from polysaccharides, and networks from various compact globular and rod-like protein particles. The order in which these systems are introduced, i.e. gels from statistical coils followed by gels from more compact and ordered biopolymers is that followed by Ferry in his 1948 Review [3], and despite the ensuing years many of the ideas discussed therein remain valid. We have therefore no hesitation in adopting what would seem to be a natural progression from gels formed by polymers in essentially random conformational states to systems which involve specific interactions between denser and less flexible particles. As will emerge; in their properties, these latter have features more in common with traditional colloidal/emulsion systems than with conventional polymer solutions.

2 Structural Techniques for Biopolymer Gels

2.1 Molecular, Macromolecular and Supramolecular Distance Scales

It should be clear from the Introductory Section of this Review that the physical techniques which have been applied to the study of biopolymer gels may be divided into those which essentially examine the details of crosslinking at the molecular level, and those which probe over much longer distances.

To clarify this, whilst spectroscopic methods, including infrared, Raman and NMR spectroscopy and chiroptical methods, (optical rotation, circular dichroism etc.) are probes operating essentially over "molecular" and slightly greater distances e.g. < 10 nm, the range from say 10–2000 nm is typically that at which small angle X-ray, and neutron, and light scattering are of major significance, and we shall refer to these as "macromolecular" probes. Thus, for gels, we could in principle, in a SAXS experiment, measure the average cross section of bundles of chains, and at longer distances, with integrated light scattering or SANS, the mean square radius of isolated chains, or dangling chain ends. Over the distance scale from 2000 nm upwards we are concerned with structural probes including electron and optical microscopy, and with mechanical measurements, which essentially monitor network continuity over macroscopic dimensions, and we shall refer to these as "supramolecular" probes. This distinction may, in some cases appear artificial, but much of the work on biopolymer gels has concentrated upon molecular studies, and from these, structural details have been inferred over macromolecular and supramolecular distances. This contrasts fundamentally with most work on covalently crosslinked synthetic gel networks, where the mechanism of crosslinking is already implied by the chemistry of crosslinking. Since, as mentioned before, some of the techniques which have been applied to examine biopolymer gels are biophysical rather than those of polymer physicists, these will be described in some detail. When, in Section 3, we discuss the mechanical characterisation of biopolymer gels, the techniques used will, on the whole, be more familiar to workers in the synthetic network area. Modifications and extensions required may be novel, however, and where this is true these will be discussed in more detail.

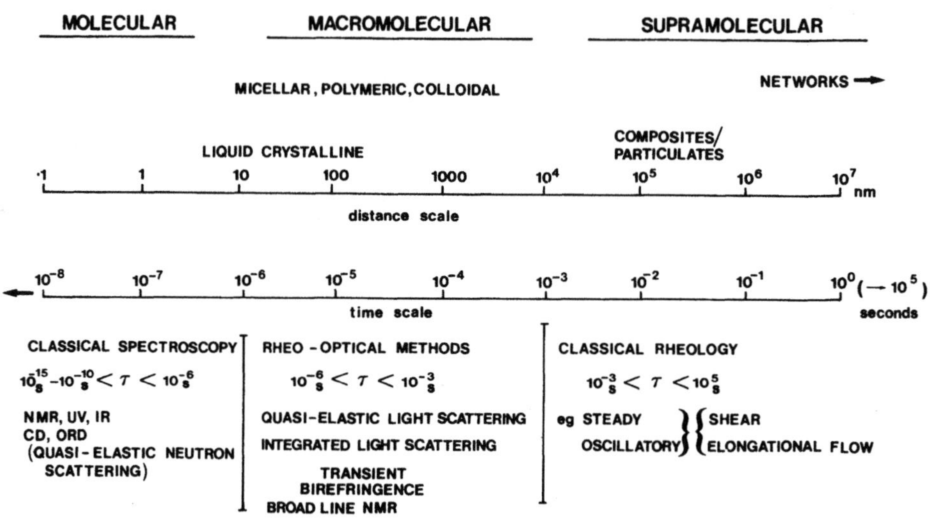

Fig. 1. Explanation of "molecular", "macromolecular" and "supramolecular" distance scales, the techniques used to probe these and the approximate time scales involved

Since the characterisation of techniques into molecular, macromolecular, and supramolecular, probes may itself be somewhat novel, Fig. 1 illustrates the concept in more detail. It relates the timescales of interest also to the distance scales, and points out where other techniques, particularly the rheo-optical methods, overlap. Whilst the latter have been widely applied to non-crosslinked biopolymers [13,14], where they have proved a sensitive and valuable tool for examining chain stiffness, studies of branched and network systems have so far been largely neglected.

2.2 Spectroscopic Methods — The Molecular Distance Scale

2.2.1 Ultraviolet Spectroscopy (UV), Circular Dichroism (CD) and Optical Rotatory Dispersion (ORD)

For biopolymers containing chromophores, measurement of radiation absorption in the UV can allow conformational changes to be monitored, and this approach [15] is particularly appropriate when polymer repeat units contain the same simple chromophore, such as the amide group in peptides and proteins, or the carboxyl group repeated in some polysaccharides. In the peptide case, conformational changes such as helix-coil transitions have been thoroughly investigated over the years by the UV approach. In conformational studies where high sensitivity is required, however, there are advantages in applying circular dichroism spectroscopy [16] or the closely related method of optical rotatory dispersion [17], and in molecular studies of protein and polysaccharide gelation these methods have been consistently favoured.

Optical rotatory dispersion and circular dichroism, applied in the UV or the visible region of the spectrum, are techniques closely related to conventional visible and UV spectroscopy. In their application, however, circularly polarised light is used, and spectra can only be obtained if an intrinsic lack of molecular symmetry prevails. In this situation left and right circularly polarised light (sometimes supplied to a sample in equal amounts in the form of plane polarised light) may be absorbed to different extents by a polymer chromophore and circular dichroism apparatus measures this difference in absorptivity as a positive or negative quantity, and expresses it as a function of wavelength. The result is a band spectrum related to the original UV absorption spectrum but containing negative, as well as positive peaks. A UV absorption band contributing to the CD spectrum in this way is described as optically active, and the size and sign of its contribution.are determined by a condition substantially different from that determining its original UV activity. Whilst the integrated intensity of a UV band depends on the square of the electric transition moment of the corresponding quantum transition, it turns out that the contribution (rotational strength) which this band makes to the CD spectrum depends on the scalar product of this quantity and the corresponding magnetic transition moment [18]. In accordance with this criterion, a given UV band may contribute positively, or negatively, in a corresponding CD experiment depending on the angle between the two transition moment vectors, and the size of its contribution is determined by the magnitudes of these.

In the study of biopolymer conformation, the particular importance of the CD effect is that a chromophore's contribution to the spectrum is sensitively determined by its molecular environment and, in particular, by the asymmetry of this environment [18]. In part, therefore, its rotational strength will depend on the configuration of the polymer residue to which it is attached, on its own orientation in relation to this configuration, and on the orientations of other residues nearby.

Applications of CD spectroscopy in the field of biopolymer gelation are common. In studies of the thermally-induced aggregation and gelation of globular proteins, for example, CD measurements have shown the extent to which native protein conformation is lost in a particular situation, as unfolding and cross-linking occur (see Sect. 5.3.1), and at times CD spectroscopy has also indicated development of new ordered conformation (for example, β-sheet) during the network forming process.

The CD method has also been extensively applied to monitor network formation by polysaccharides, study of the alginates, seaweed polysaccharides from brown algae, having produced particularly interesting results. Solutions of these polymers in their sodium salt form can be gelled by the introduction of calcium ions (Sect. 4.2.1), and changes in the rotational strengths and positions of CD bands associated with the alginate carboxyl groups have led to a molecular description of net work formation. In this so-called "egg-box" model, segments of the alginate polymer chain associate to form ordered dimeric structures with the calcium ions packed into electronegative cavities like eggs in an egg-box.

Unfortunately, the CD approach, powerful as it is, is not universally applicable to situations where biopolymers gel. Apart from obvious difficulties, arising when samples have no chromophores or are extremely turbid, there is a technical barrier imposed by the nature of most present-day instrumentation. Most conventional spectropolarimeters cannot measure CD spectra accurately below ~190 nm and this means that polymers having chromophores absorbing below this limit cannot be studied. Though recently developed vacuum UV CD instruments can probe to lower wavelengths, and have been used in some gel studies, these are not so far routinely available, and have not yet substantially reduced the problem.

The absence of a measurable CD spectrum for a polymer has not deterred workers in the biopolymer gel field in their efforts to apply chiral methods. In this situation they often resort to the intimately related technique of optical rotatory dispersion [17], that is the measurement of optical rotation over a range of wavelengths. Although the ORD spectrum contains no new information, being directly related to the CD spectrum by a mathematical transform (Kronig-Kramers) [18], it presents the CD data in a quite different form, the optical rotation at one wavelength (in the visible, for example) being often determined by several CD bands some distance away in wavelength. It follows that inaccessible CD bands can sometimes be monitored indirectly by ORD measurements though, of course, disentangling contributions, if more than one band has an influence, can be tricky. The relationship between a typical CD band and its contribution to the ORD spectrum is indicated in Fig. 2.

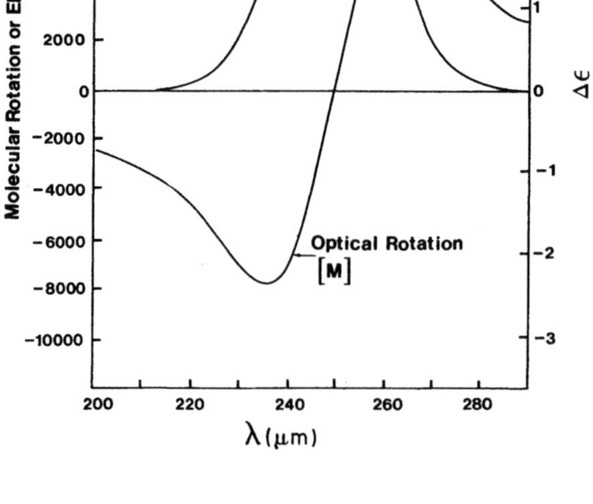

Fig. 2. Model Gaussian CD band (centred on $\lambda = 250$ mμ) and corresponding optical rotatory dispersion curve. Reprinted with Permission from Accounts of Chemical Research, *1*, 144, (1968), Fig. 1, p. 145. Copyright 1968 American Chemical Society

Optical rotation measurements have commonly been made on aggregating or gelling biopolymer systems. The method has been widely applied to study gelatin gelation (Sect. 4.1), for example, and it has featured prominently in studies of the gelling of seaweed polysaccharides such as agar and the carrageenans (Sect. 4.2.1). Agar and the carrageenans do not have easily accessible CD bands (unlike the alginates) and early conformational studies of these polymers in solution relied upon observations of optical rotation at visible wavelengths. By extensive studies of the optical rotatory powers of polysaccharides in general, in relation to their conformational preferences, Rees established a semi-empirical relationship between torsion angles specifying the relative orientations of polysaccharide residues in polysaccharide polymers and the magnitude and sign of the optical rotation. Applying this relationship to data for cold-set gelled solutions of carrageenans, and, in particular, to optical rotation-versus-temperature data, he and his co-workers established a model for the cross-linking process (Sect. 4.2.1). In this, cross-links were described as co-operatively formed "junction zones" arising as segments of the carrageenan polymers combined to form ordered double helices (and in some cases clusters of these helices). Co-operativity was inferred from the suddenness of the optical rotation change in relation to temperature, and the presence of the helix was deduced by combining the lowest temperature optical rotation result with fibre diffraction data for the double helix, and applying the semi-empirical formula.

It is interesting that in the agar and carrageenan cases justification for the optical rotation argument has come recently from the results of vacuum ultraviolet CD investigations. These have shown that these polymers give rise to at least two conformationally sensitive bands below 190 nm and Kronig-Kramers transformation has confirmed the strong influence of these bands on the optical rotation at longer wavelengths [19].

2.2.2 Infrared (IR) and Raman Spectroscopy

These techniques measure spectroscopic transitions which have their origins in changes in molecular vibrational energy [20], but whilst IR is a conventional absorption spectroscopic approach, Raman spectroscopy is based on the inelastic scattering of visible light. The two approaches are complementary in fact since, whilst they both monitor vibrational transitions, they are subject to different selection rules; Raman spectroscopy monitoring vibrations associated with a change in molecular polarizability, and IR spectroscopy vibrations associated with a change in electric dipole moment.

Another distinction between the IR and Raman approaches is that, whilst water absorption tends to obscure the IR spectra of biopolymer solutions and gels, water is essentially invisible in the Raman case, and in the Raman experiment a great many more spectral details can be measured. In part, this disadvantage of the IR technique has been reduced by the introduction of Fourier transform IR spectrometers, but problems still arise, particularly where dilute systems are of interest.

IR and Raman spectroscopy have both contributed to molecular studies of biopolymer gelation. This is particularly true for gels obtained from globular proteins by heating (see Sect. 5.3.1), as in CD studies of these systems, vibrational spectroscopy has proved to be a viable way of monitoring loss of native conformation during aggregation and the development (in some cases) of new ordered structures (β-sheet conformation, for example). This is because the modes of vibration associated with the peptide backbone, and in particular those which correspond to stretching of the carbonyl group, are conformationally sensitive, both in terms of frequency and intensity [21–23]. Again complementarity is found between the IR and Raman methods, since IR is best suited to monitoring changes in β-sheet conformation (see Fig. 3) whilst Raman spectroscopy allows the α-helix content of proteins to be followed separately [24] from β-sheet and random coil (see Fig. 4). IR and Raman spectroscopy, together with CD, are often applied in concert in studies of protein gelation, but the

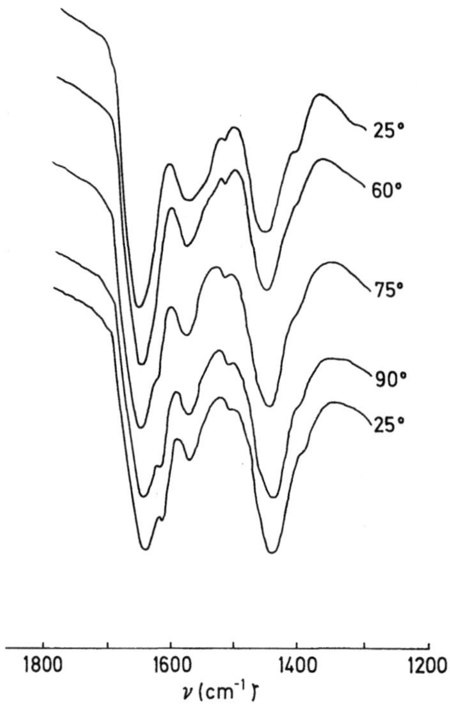

Fig. 3. Infrared absorption spectra measured during heat-set gelation of a typical globular protein solution (bovine serum albumin, 10% w/w in D_2O). The development of a shoulder (1620 cm^{-1}) in the Amide I carbonyl stretching band at ~1650 cm^{-1} indicates partial reorganisation of the peptide structure to β-sheet. Reprinted with Permission from International Journal of Peptide and Protein Research *17*, 353 (1981), Fig. 1, p. 355. Copyright © 1981 Munksgaard International Publishers Ltd., Copenhagen, Denmark

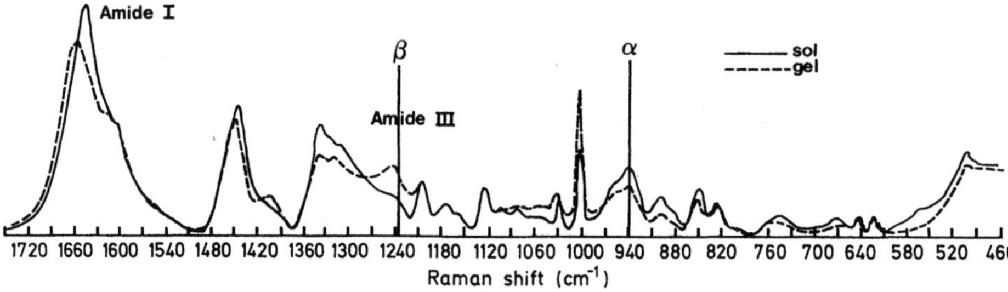

Fig. 4. Comparison of sol and gel laser Raman spectra for system described in Fig. 3. Changes in α-helix and β-sheet secondary structure contents may be inferred from bands at ~940 and 1240 cm^{-1} as shown. Reprinted with Permission from "Functional Properties of Food Macromolecules", edited by J. R. Mitchell and D. A. Ledward, Fig. 23, p. 250, Elsevier Applied Science, 1986

two vibrational technique are generally more productive as they are less affected by sample turbidity.

Where polysaccharide network formation is concerned, the IR and Raman methods have had much less impact. This is probably because bands strongly sensitive to conformational change are less easy to identify than in the protein and peptide situation. IR methods have been employed in the analysis of residue content of polysaccharide samples, however, and in the study of ion binding to sulphate groups (Sect. 4.2.1).

2.2.3 Nuclear Magnetic Resonance (NMR) Spectroscopy

NMR methods have been applied to the study of biopolymer gelation, particularly to polysaccharides [25–32], but also to gelatin [33]. The method has not played a key role, however, in the identification of molecular cross-linking mechanisms or gel structure, but recent studies of cation binding in carrageenan gels (see Sect. 4.2.1) by cation NMR [31] represent a move in this direction. In addition, it seems certain that structural information about biopolymer gels can be derived from NMR studies of water (solvent) proton relaxation times [32, 33].

The rather low involvement of the versatile NMR technique in biopolymer gel studies probably stems from the nature of the gelation process itself. In most cases this involves chain rigidification and immobilisation and this in turn usually leads to a loss of high resolution NMR signal. Thus, the NMR experiment is more able to measure the structural and dynamic characteristics of the biopolymer in solution, prior to aggregation, than it is to describe the network which is formed later.

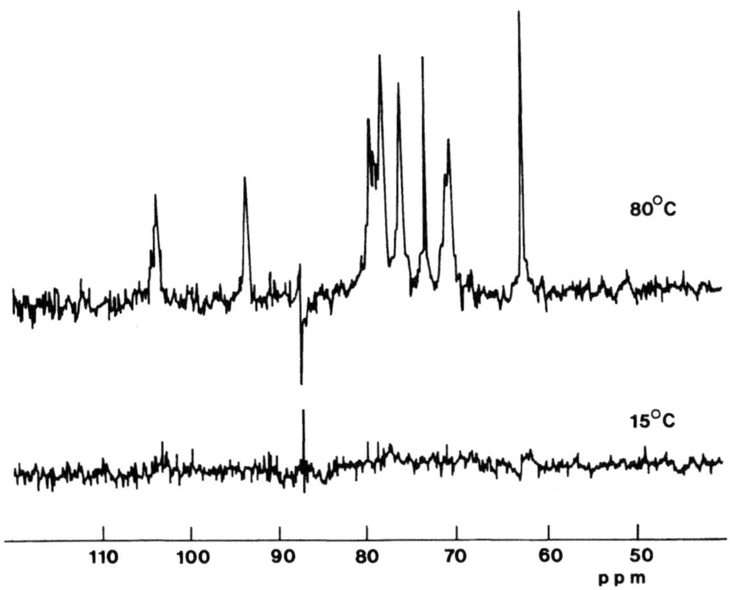

Fig. 5. Loss of high resolution ^{13}C-NMR spectrum as a solution of ι-carrageenan segments (6% w/v in 0.1 M NaCl) is cooled. Reproduced with Permission from Faraday Discuss. Chem. Soc. *57*, 221 (1974), Fig. 3, p. 225

Despite the above limitations in the gel context, nevertheless, NMR spectroscopy has played a role in studies of polysaccharide network formation. It has allowed the cooperative nature of conformational changes taking place as carrageenan gels form to be clearly demonstrated, and has indicated the very substantial rigidification of carrageenan chains which accompanies double helix formation (Fig. 5). In this work [25] ^{13}C spectra were measured as, prior to aggregation of the carrageenan, these were better resolved than were alternative proton spectra.

High resolution NMR studies of chain immobilisation are intrinsically limited in the information they provide, for, as network formation proceeds, the presence of network is inferred negatively by the absence of an NMR signal, rather than positively, by a direct measure of a spectrum from the immobilised moiety. In some cases, however, network formation by a biopolymer may involve only a limited immobilisation of chains (e.g. entanglement network) and here NMR methods may be used to monitor all of the network-forming material.

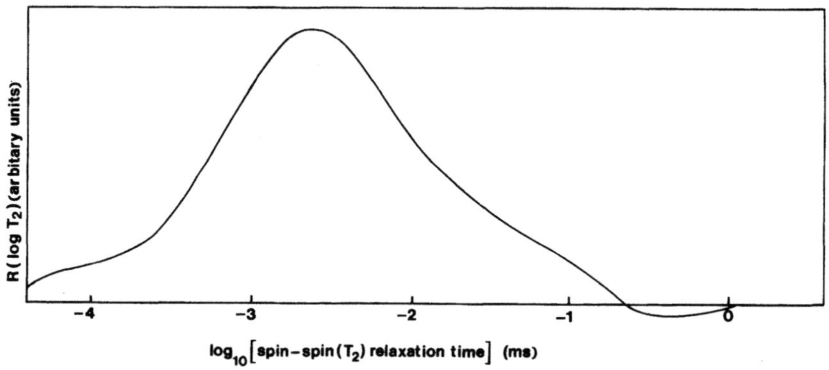

Fig. 6. Proton NMR spin echo decay function for sodium hyaluronate (5% w/v in D_2O: 353 K) deconvoluted to yield transverse relaxation time (T_2) distribution function. Reproduced with Permission from Macromolecules *15*, 587 (1982), Fig. 7, p. 601. Copyright 1982 American Chemical Society

Ablett et al. [26] for example, have combined high resolution proton NMR studies of certain polysaccharides in D_2O, with broad line NMR data obtained using a spin-echo spectrometer. This latter technique, allowed the transverse relaxation times of all protons in the polymer to be monitored, albeit at the expense of chemical shift information lost in the spin-echo experiment. For polysaccharide chains restricted in their motion, the transverse relaxation times (T_2) were assumed to be broadly distributed, and a Fourier transform approach was employed to extract T_2 distributions from spin-echo decay profiles. This method was found to be particularly successful for hyaluronic acid [34] (a polysaccharide component of cartilage and connective tissue — Sect. 4.2.4) since the highly viscous solution formed at neutral pH by this polymer was shown to have a broad proton T_2 distribution (Fig. 6) consistent with the non Lorentzian character of absorption peaks in its high resolution proton NMR spectrum, and indicative of an entanglement network composed of stiffened hyaluronic acid polymer chains.

2.3 Scattering and Diffraction Methods — The Macromolecular Distance Scale

2.3.1 X-Ray and Neutron Diffraction/Scattering

Scattering and diffraction methods, using X-rays [35] and neutrons [34], can provide structural information about biopolymer solutions and gels over both the molecular distance scales referred to previously, and over the longer distance scales which were earlier called macromolecular, the probing distance being approximately the inverse of the wave vector q (cf. Fig. 1).

At short range, wide-angle X-ray diffraction methods have provided indirect evidence about the molecular structures in biopolymer gels for, although the gels themselves cannot easily be studied by this method on account of low network scattering, diffraction investigations of fibres and dried gels, prepared from the hydrated materials, have proved highly informative. Such solid state structural information, unless supported by more direct forms of evidence, must of course be regarded as extremely tentative, but the structural models produced provide useful starting points in searches for network descriptions. In some situations, such as carrageenan gelation, for example, the fibre diffraction structure has become established as predominating in the more fully hydrated state (Sect. 4.2.1) whilst in others, such as gels from denatured globular proteins (Sect. 5.3.1) the situation is more complicated, and early ideas from fibre diffraction have been misleading. Even in this last case, however, evidence from wide-angle diffraction has proved useful, for comparisons of powder patterns [35] for dried globular protein solutions and gels have given val-

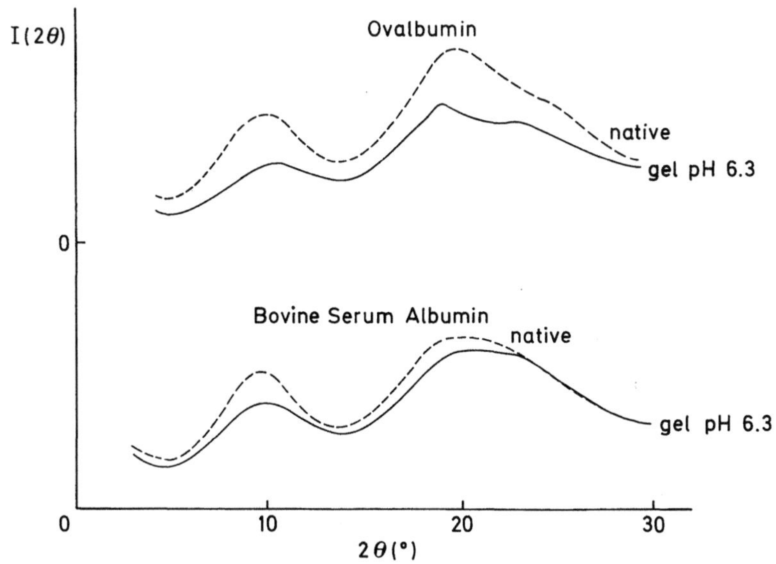

Fig. 7. Comparison of wide-angle X-ray diffraction data (I(2θ) versus 2θ:λ = 1.542 Å) for dried and powdered sols and gels from the globular proteins, ovalbumin and bovine serum albumin. The ovalbumin gel forms with a greater loss of native protein conformation and a greater shift towards β-sheet secondary structure (peak at 2θ ∼ 19°)

uable evidence about the extents to which such proteins lose their native corpuscular forms (tertiary and secondary structures) on aggregate formation (Fig. 7).

Studies of biopolymer solutions and gels over macromolecular distance scales have also been performed, principally using small-angle X-ray scattering techniques [35], but some neutron data is available [34]. For the gels based on denatured globular proteins, scattering from the networks at small angles is strong (Fig. 8), and can easily be monitored [35], as can changes in scattering accompanying sol-gel transformations. For polysaccharide gels, however, the situation is more difficult experimentally, as the network scattering is weak, and great care is required if it is to be accurately separated from the solvent contribution. High accuracy is desirable, however, for in the polysaccharide case, the tails of the scattering curves, if measured precisely, have the potential to provide estimates of the cross-sectional dimensions of chains present in both sol and gel, and to provide critical measures of the scattering characteristics of ordered species possibly present such as helices, chain dimers,

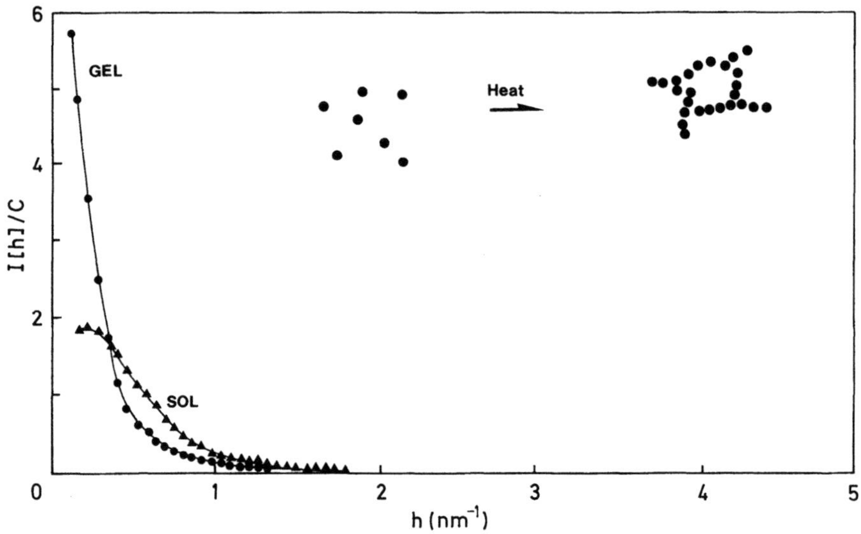

Fig. 8. SAXRS data (scaled for concentration) for typical globular protein sol and heat-set gel ($h = 4\pi \sin \theta/\lambda$, $\lambda = 1.542$ Å). Reprinted with Permission from "Functional Properties of Food Macromolecules", edited by J. R., Mitchell and D. A. Ledward, Fig. 26, p. 259, Elsevier Applied Science, 1986

clumps of helices etc. In view of the difficulties and uncertainties attached to the study of polysaccharide gelation by the small-angle technique, few results have been reported, but data presented recently by Clark and Lee-Tuffnell [35] for lithium ι-carrageenan appears to be consistent with the classical double helix description (Fig. 9). Neutron scattering data [34] for aggregated ι-carrageenan apparently also supports this model. Neither technique, it should be added, can realistically exclude other forms of chain association if these can provide rod-like structures of similar dimensions.

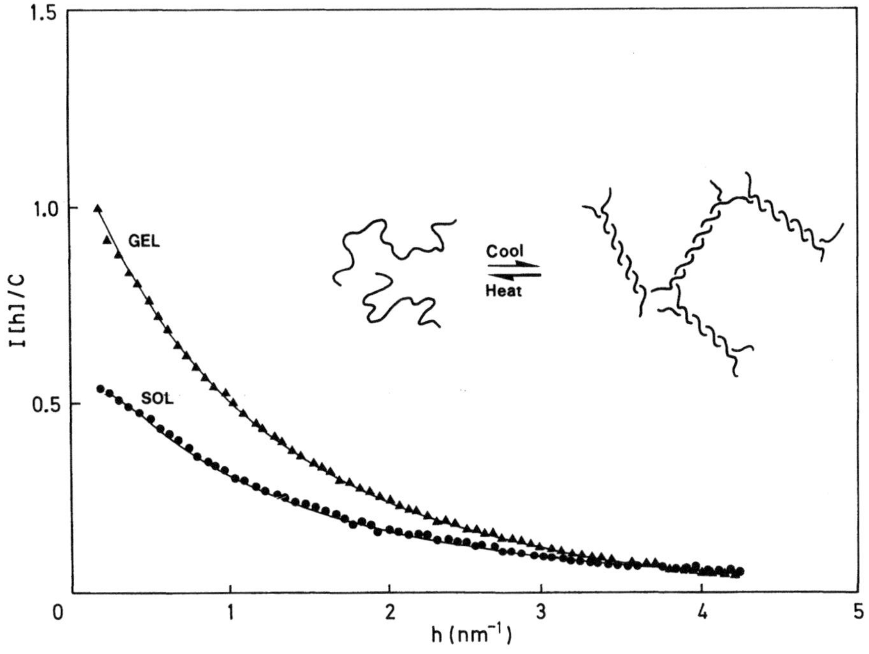

Fig. 9. SAXRS data (scaled for concentration) for sol and cold-set gel from lithium ι-carrageenan (2% w/w). Reprinted with Permission from "Functional Properties of Food Macromolecules", edited by J. R. Mitchell and D. A. Ledward, Fig. 27, p. 260, Elsevier Applied Science, 1986

In the globular protein gel case, whilst experimental measurements are easy to obtain, theoretical interpretation is confused by the combination of unfolding and aggregation involved, and radial distribution approaches, based on a corpuscular description of such networks, have met with only limited success (Sect. 5.3.1). In general, however, small-angle X-ray scattering studies of heat-set globular protein gels have supported the corpuscular hypothesis, and have demonstrated how fundamentally different these networks are from the fine fibrous assemblies characteristic of gelatin and most polysaccharide gels.

In the context of globular protein gels and their molecular description, it is worth noting that the formation of aggregates and networks from associating particles is currently a topic of great interest in pure physics and one which is being extensively examined theoretically. Recent studies have confirmed, for example, that aggregation mechanisms such as f-functional random polycondensation and diffusion-limited Smoluchowski kinetics are just special examples of a more general coagulation mechanism [36]. Considerable attention has been paid to characterising the structures formed, for example, in computer "clumping" experiments in terms of the Hausdorff or fractal dimensions [37]. If N is the number of units which occur within a sphere of radius R, then the Hausdorff dimension is defined by $N \sim R^D$. Whilst this approach is currently of immense interest to physicists [38], it is more of peripheral interest in the context of the present article which is limited to gel systems. However, it is worth pointing out the work of Witten [39], who used computer simulation to treat the case of diffusion-limited aggregation. By calculating the density correlation function (DCF) defined by,

$$G(r) = \langle \varrho(o) \cdot \varrho(r) \rangle \tag{1}$$

with $\varrho(r)$ the density of particles in a volume element in space r from an origin o, he observed that for structures obeying a law of the form,

$$G(r) \propto r^{-A} \tag{2}$$

$N \propto R^{3-A}$, i.e. $D \propto 3 - A$. The "structure" of the aggregated species can then be defined in terms of the single parameter D. It has been suggested that the aggregation behaviour of the globular proteins referred to above could be described in this way [40], and differences in structural characteristics induced by changing gelling conditions (temperature, pH, ionic strength) could no doubt be succinctly characterised in terms of D. The X-ray scattering approach described above could make a contribution to this measurement procedure particularly when coupled with longer-range information from the light scattering and microscopical techniques to be described in the next two Sections.

2.3.2 Light Scattering (Static and Dynamic)

The use of light scattering to study biopolymer aggregation, and to provide a long-ranging structural description of gels is an attractive prospect, and today it is made even more interesting by the fact that the procedures of classical (static) light scattering can be supplemented by those of the more recent dynamic approach.

Early studies, however, involved static light scattering alone and in these, applications of light scattering to aggregating protein and polysaccharide systems were aimed largely at molecular weight determination. For example, the doubling of molecular weight, which occurs on cooling certain ion forms of degraded (segmented) iota carrageenan polymers (Sect. 4.2.1) was first reported from light scattering experiments, and today, light scattering is still used to provide details of aggregation in polysaccharide systems [34].

In the protein field [35], light scattering has been used by Kratochvíl and co-workers to measure molecular weight — time profiles for globular proteins induced to aggregate by heating (Sect. 5.3.1) and a branching model applied to interpret results, and in the area of milk proteins, the aggregation of casein micelles (Sect. 5.4.2) induced by enzyme action, has also been followed by this technique (mainly turbidity measurements). In this latter case, a Smoluchowski mechanism was invoked to describe the association event.

Applications of light scattering, in a conventional sense, to follow molecular weight changes are useful, but recently, more sophisticated treatments of biopolymer aggregation, using light scattering, have emerged. In these, the angular dependence of scattering has been employed to provide structural information, much as small-angle X-ray scattering can be applied to somewhat smaller particles.

An example of this more structural application of light scattering is the study of the polymerisation of fibrin [34] to form the fibrin network, or clot, which is the basis of blood clotting (Sect. 5.2.5). Here, by plotting light scattered intensity — versus — scattering angle data in various ways (Zimm plot, Holtzer plot, Kratky plot) and for data sets obtained at various times during aggregation, Burchard and Muller were able to describe the event in terms of an initial linear association of rod-like fibrin monomers, followed by occasional branching, and under certain circumstances, lateral association of the linear elements. In this work, light scattering experiments also provided valuable information about the structure of the fibrinogen precursor of fibrin, this being shown to be basically a rod, but to be capable of existing in solution in more than one structural form.

The above study of fibrin and fibrinogen, also included application of dynamic light scattering. Differences in translational diffusion coefficients for human and bovine fibrinogen determined by this technique, indicated that, on average, a longer rod was present in the human fibrinogen case, whilst studies of fibrin polymers of various sizes indicated hydrodynamic behaviour corresponding to ellipsoids, long rods, and branched (flexible) polymers based on rod elements.

These conclusions were drawn from plots of D_{app}/D versus $q^2\langle S^2\rangle_z$ (Fig. 10) where D_{app} is the apparent Z-average translational diffusion coefficient, D is the true Z-average diffusion coefficient, $q = (4\pi/\lambda) \sin\theta$ is the scattering vector, and $\langle S^2\rangle_Z$ is the Z-average square of the radius of gyration.

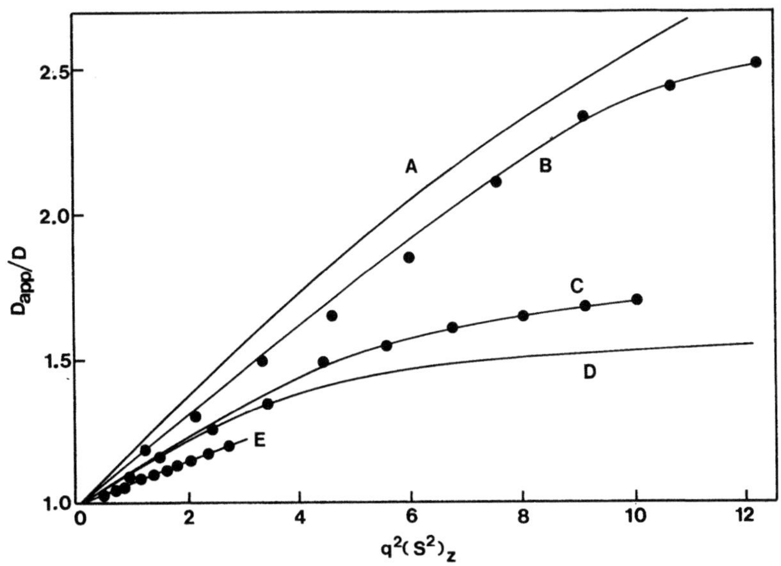

Fig. 10. Translational diffusion coefficient data (D_{app} versus $q = 4\pi \sin\theta/\lambda$ = X-ray h) for fibrin polymers of increasing molecular weight (C → B) plotted in the reduced form D_{app}/D_z versus $q^2\langle S^2\rangle_z$, where D_z and $\langle S^2\rangle_z$ are the z-average translational diffusion coefficient and mean square radius of gyration respectively. Data for polymers of a given shape lie on common lines, and curves D and A represent extremes of behaviour expected for thin extended rods and random coils. Data for fibrin polymers of lower molecular weight (E) and having ellipsoidal shapes fall below line D. Reproduced from Müller, M. and Burchard, W. International Journal of Biological Macromolecules, 1981, 3, 71–76, by permission of the publishers, Butterworths & Co. (Publishers) Ltd. ©

The technique of quasi-elastic light scattering has been applied by a number of groups to the study of biopolymer gels, and at this point it is appropriate to consider the behaviour expected from both permanently cross-linked and entanglement networks. The latter have crosslinks of finite lifetime T_τ, but for frequencies $\omega \gg T_\tau^{-1}$ (cf. Sects. 3.2.4 and 3.3) they have been termed pseudogels [41]. Experiments by Munch et al. [42] on both types of system have demonstrated that measured diffusion coefficients for crosslinked polymer networks, and for concentrated (strictly semi-dilute) solutions, show very nearly the same increase in effective diffusion coefficient D_Z with concentration (when the same polymer/solvent pair are selected), and that this $D_Z \propto C^{0.7}$, which compares with the scaling prediction $D_Z \propto C^{0.75}$. It is important to appreciate that the values of D_Z obtained by Munch et al. were derived carefully, by extrapolation to $q = 0$, from measurements over a range of q. This is crucial since, in general, we would not expect apparent diffusion coefficients determined at finite angles of scatter (for example $\theta = 90°$) to be as physically meaningful as the extrapolated results. It should be added that, whereas in much of the early literature on biopolymer gels "single exponential" behaviour has been reported, more comprehensive modern studies involving a range of scattering angles, often reveal complex decay characteristics and invoke the need for careful multiexponential deconvolution.

The mutual diffusion coefficient obtained above may be related to the frictional resistance of the sample by,

$$D_Z = C(\partial\Pi/\partial C)_T\,(1 - \bar{V}_2 C)/f \tag{3}$$

where $(\partial\Pi/\partial C)_T$ is the osmotic compressibility, f is the friction coefficient and $\bar{V}_2$ is the polymer partial specific volume. Since $C(\partial\Pi/\partial C)_T$ has the units of pressure (a modulus) the product $D_Z f/(1 - \bar{V}_2 C)$ has been termed the "osmotic modulus" M_{os} [43–45]. Results of Candau et al. [44] demonstrate that data for M_{os} follow approximately the same power law dependence on crosslinking density as mechanical data, although absolute values are apparently rather different, since the light scattering approach is probing longitudinal modes involving concentration fluctuations (and presumably also high frequencies).

As mentioned previously, complex relaxation behaviour is normally observed in the case of semi-dilute solutions and gels, and in practice, an additional slow mode is often detected in the time correlation function, which increases in amplitude as the concentration increases (Fig. 11) [46–48]. For the entangled solutions this effect is usually attributed to the self-diffusion of chains, and according to de Gennes [41] should follow a scaling law,

$$D_{self} \propto C^{-7/4} \tag{4}$$

In permanently crosslinked gels, by contrast, there is the so-called co-operative diffusion of chain segments having a correlation length ζ (or "mesh size"), but no self-diffusion. Data for biopolymer gels is again not so plentiful, but observations by ter Meer for ι-carrageenans [49] (Sect. 4.2.1) are particularly interesting. In this, case self-diffusion behaviour is seen both above and *below* the sol/gel transition, but at low temperatures $D_{self} \propto C^{-\tau}$, with $\tau \sim 4$–5. Only at higher temperatures, when the gel

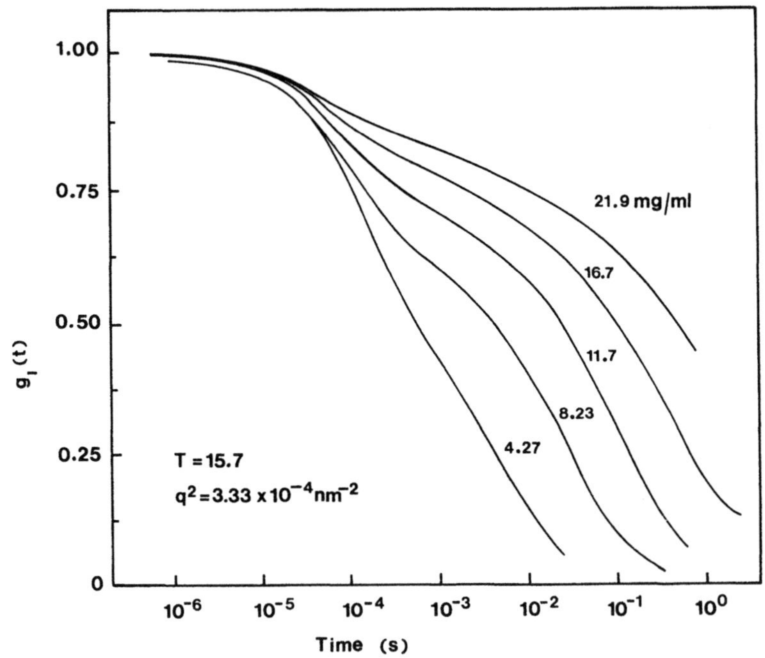

Fig. 11. Time correlation function of scattered light for gels of ι-carrageenan at different concentrations. Two relaxation processes are clearly seen. Reproduced with Permission from "Integration of Fundamental Polymer Science and Technology" edited by L. A. Kleintjens and P. J. Lemstra, Fig. 3, p. 232, Elsevier Applied Science, 1986

has "melted out", does τ decrease to ~2–2.5 and so begin to approach the value expected for entanglement network systems. These data strongly suggest that junction zones in the ι-carrageenan network can open and reform allowing the chains to move on the timescale of the experiment. Application of the modern generation of logarithmic correlators [50] should be of considerable value in future studies of such phenomena.

A slightly different technique to QELS is forced Rayleigh scattering (FRS) [51] in which a small proportion of a polymer sample is labelled photochromically, and then the movement of the labelled species followed in a matrix of the unlabelled material. Data for gelatin by Chang and Yu, using FRS [52], and by Amis et al. [47] using QELS, also seem to give $\tau > 7/4$. However, in view of the controversy regarding the nature of the slow mode in synthetic polymer solutions, this whole area is still somewhat in its infancy, and much future work is to be anticipated.

Other types of investigation of biopolymer gels by dynamic light scattering have been reported, such as work by V. J. Morris on ι-carrageenan, in which rigidity moduli were determined by the monitoring of gel vibrations and plotted against concentration [53]. Other systems investigated have included alginates and agarose (Sect. 4.2.1), an early study of gels from agarose having been made by Prins and coworkers [54,55]. From the results of both static and dynamic light scattering experiments they concluded that a Cahn-Hilliard spinodal decomposition mechanism (Sect. 3.7) was exhibited. The time correlation function also showed strongly damped oscillatory behaviour, at least for some hours after cooling below the sol/gel transition. This may be attributed to mass flow accompanying the gelation. Similar behaviour has been observed by Sellen and coworkers for agarose-water systems [56].

2.4 Microscopical Methods — The Supramolecular Distance Scale

Light and electron microscopy are techniques which have long been used to probe biopolymer gel structures over long distances. The main drawback, however, is that gel samples, which are often largely water, must be extensively processed before

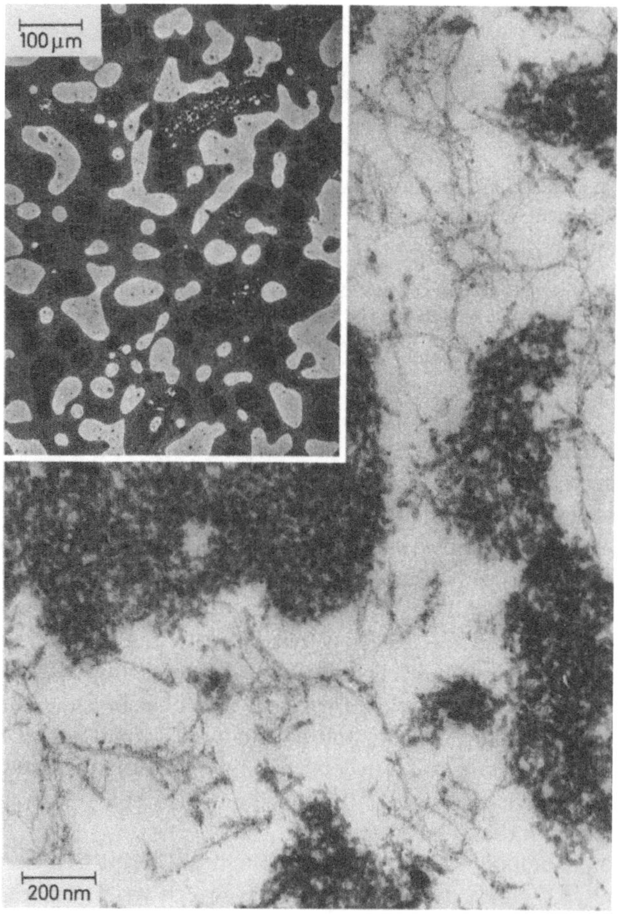

Fig. 12. Transmission electron and light (insert) microscope images for a mixed thermally-set biopolymer gel based on bovine serum albumin (globular protein) and agarose (polysaccharide). In the electron micrograph, the protein network appears as the denser, more darkly staining, phase. The light micrograph reveals a highly complex microstructure containing the mixed gel phase of the electron micrograph and predominantly protein and polysaccharide containing phases (very dark and light areas, respectively)

micrographs can be obtained. The processing procedures necessary vary, depending on what kind of microscopy is involved, but they are usually drastic. In transmission electron microscope studies of gel sections based on gelatin, (Sect. 4.1.2), polysaccharides (Sect. 4.2.1), and denatured globular proteins (Sect. 5.3.1), for example, the native gels must be subjected to fixing, staining, dehydration and embedding procedures, and hence there is no guarantee that micrographs finally obtained will give any indication of the true microstructures involved. Corroborative evidence is always required [35], either from the structural approaches already discussed, or from other independent microscopical approaches (e.g. freeze-fracture, scanning electron microscope, etc).

For polysaccharide gels, little corroboration has been achieved for the faint wispy networks (Fig. 12) which are usually indicated, and the same applies to gelatin. For the denatured protein gels, however, a good case has been made for the reliability of transmission electron microscope images (see also Fig. 12), this having been established by careful comparative studies [35] involving the application of several different microscopical approaches, and other techniques such as wide-angle and small-angle X-ray scattering.

Although not a major part of the present article, it is interesting to note that where gels containing more than one biopolymer component have been studied, a combination of light and electron microscopy has generally proved valuable. In many cases a microphase-separated microstructure (Fig. 12) has been observed [35] which bears a strong resemblance to published micrographs of interpenetrating networks of synthetic polymers. Despite this we must issue the conventional *caveat* about generation of artefacts.

2.5 Thermodynamic and Kinetic Methods

Because of the non-equilibrium character of most biopolymer gels, thermodynamic approaches might seem unlikely sources of useful information about gels and gelling processes. Nonetheless, methods such as differential scanning calorimetry, osmometry, equilibrium dialysis, permeability and swelling experiments, and ion activity measurements have been applied. In addition, statistical mechanical theories such as those describing co-operative conformational changes (Sect. 3.5), or ion binding, have sometimes been invoked to interpret experimental observations. A recent attempt [35] to relate aspects of network structure to rheological behaviour has also invoked equilibrium thermodynamic concepts as did early theories of gel melting behaviour applied to gelatin (Sect. 3.5).

In the polysaccharide area (Sect. 4.2), application of the above methods has been particularly widespread. Thus, quite early in the study of carrageenan aggregation, a combination of differential scanning calorimetry, optical rotation measurements, and helix-coil transition theory was employed to study the conformational change occurring in carrageenan samples on cooling. A description of the process as a fully co-operative dimerisation of carrageenan chains via double-helix formation was achieved, provided that the influence of chain polydispersity was properly introduced in calculations. In more recent studies of carrageenan aggregation, the totally co-operative description has been challenged, but again calorimetric and optical rotation methods, coupled to theory, have provided the backbone of investigations. In this more recent work, the experimental approach has been strengthened by the introduction (particularly for the carrageenans, but also for xanthan) of fast kinetic measurement techniques such as stopped-flow polarimetry. This last approach, in the form of salt-jump experiments, has provided still more evidence in support of the double-helix model for carrageenan cross-linking. Calorimetric and kinetic optical rotation approaches have also been employed in the study of gelatin gelation (Sect. 4.1) and in this case the rather slow change of optical rotation with time, at a particular temperature, has been analysed to provide a kinetic description of triple helix formation, and the further growth and association of triple helices at longer times.

Other approaches, which may be described as thermodynamic have been applied in biopolymer gel studies, particularly to polysaccharides. Studies of alginate gelation, for example, involving ion binding measurements (activity measurements and equilibrium dialysis) have been used to support the "egg box" model for cross-linking discussed in Sect. 4.2.1 — and have demonstrated the highly co-operative character of this process, and the essentially dimeric nature of the junction zones. In addition, swelling experiments, also performed on alginate gels, and coupled with rheological data, have provided evidence about the non-Gaussian nature of network chains between junction zones in these systems. This aspect of gel network structure, which is currently of great interest, will be mentioned again in the next Section, which deals with the mechanical characterisation of biopolymer gels, and the sol-gel transition.

3 Mechanical Characterisation and Properties of Biopolymer Gels

3.1 Introduction to Mechanical Characterisation

The experimental techniques which have been applied to investigate the rheological properties of biopolymer gels may, in principle, be divided into small-deformation measurements, used to probe viscoelastic properties within the linear regime, and large-deformation measurements designed to obtain full stress-strain profiles and failure properties. In practice, the number of small-deformation studies available is limited (although this is already an expanding area) and the majority of rheological results reported so far have been from essentially empirical failure studies. These have mainly involved the compression of samples under constant load, or penetration using a conical probe, and because of the limitations of such experiments, and the difficulty of defining the test geometry, true stress-strain profiles have rarely been obtained. Rather, such parameters as "gel strength" and "peak stress" appear quite widely in the biophysical literature [57-59]. Whilst such measurements are described, where relevant, in the Sections of this Review covering "systems", the present Section will concentrate upon more rigorous experimental approaches and on the success (or otherwise) of theory to relate results of these to the fundamental nature of the systems investigated. Possibilities for relating network structure to properties at this more basic level by performing characterisations of pre-gel and critically branched materials will also be discussed. Whilst the importance of this approach is well appreciated by workers in the synthetic polymer area, only recently has the same been true for workers in the biopolymer area. In fact, one might argue that since the mechanism of crosslinking itself is not always completely understood, such studies are even more crucial for understanding biopolymer systems.

3.2 Small-Deformation Rheological Measurements

3.2.1 Gel Cure Experiments

Since the majority of aggregation processes to be discussed will involve physical changes (for example, of temperature) for their initiation a typical regime of mechanical characterisation would be to dissolve a biopolymer in water (electrolyte), at a temperature quite different from that at which gelation can occur, and then either to change the temperature at a reproducible rate, or to maintain the system isothermally at a new temperature at which aggregation can proceed. The system is then monitored until the gel is formed. Obviously the ideal experiment would involve small-strain oscillatory shear measurements as a function of time, at constant low frequency. Such investigations have been performed by a number of workers [60–62], and with modern equipment the procedure is largely routine, a typical trace being illustrated in Fig. 13. It should be added, however, that there are a number of points to be appreciated in this type of experiment. Firstly the frequency of oscillation, ω, cannot be too low for practical reasons, and ω ~ 1 Hz would appear a reasonable compromise

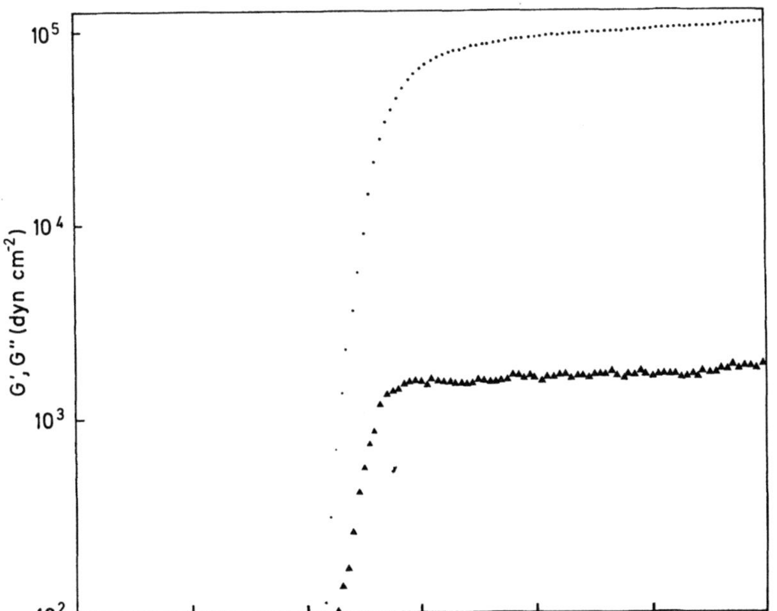

Fig. 13. Gel cure trace, showing the increase in G′ (●) and G″ (▲) for a typical heat set gel (10% w/w BSA, Ref. [71])

between measuring so fast that entanglements are included and so slow that not enough data is obtained [63]. However one must be aware that the frequency of measurement should strictly be less than the reciprocal of the longest relaxation time of the gelling material [11], and this means that the frequency should decrease and become zero at the gel point. Practically, this approach is never adopted since more, rather than less, data is needed close to the gel point. Secondly, the oscillatory strain must always be kept sufficiently low that all measurements are taken within the linear viscoelastic regime, and again this defeats the possibility of performing the experiment rigorously. There are two reasons for this; firstly at, and close to, the gel point even infinitesimally small strain will in all probability, modify the network continuity for a system in which only physical crosslinks are involved and secondly, since in this region it is likely that strain and frequency are coupled in a non-trivial manner, we should ideally perform the experiment at both zero frequency and at zero strain. For the pragmatist, however, we describe the results obtained in a typical experiment.

In Fig. 13, the real G' (storage) and imaginary G'' (loss) parts of the dynamic shear modulus are charted as functions of time. These traces have the expected shape, with G' increasing rapidly at first and then more slowly, and $G'' < G'$. In experiments of this type carried out for synthetic polymers, G'' usually passes through a maximum some time after the gel point, and then decreases again [64]. This effect has been related by Valles [65] to the contribution from dangling chain ends and from the sol fraction. In some biopolymer systems (e.g. gelatin) this same behaviour is observed [66], whereas for others it is not seen. This observation is consistent with the Valles argument, allowing for the effect of chain stiffness, and also the rapid decrease of the sol fraction of a low functionality system (see later).

Djabourov and co-workers [67] have carried out a systematic study of the above kind for gelatin gels, different temperatures being employed and a Weissenberg rheogoniometer being used for measurement. Very similar experiments have been carried out by McIntire and co-workers for systems including solutions of the muscle protein actin and blood platelet extracts [68,69], and by ourselves for a range of systems including gelatin, various globular proteins, polysaccharides and mixed gels of protein/protein and protein/polysaccharide components [66,70,71].

As far as pre-gel studies are concerned most such measurements have been carried out by scattering or spectroscopy. However the work by Peniche-Covas et al. [60] for gelatin illustrates a systematic approach for characterising a system through the gel point. They employed a sphere moving in a magnetic field within the sample and, before the gel point, the system could be measured as a viscous fluid by using a direct current and following the movement of the sphere optically. After the gel point an oscillating current was applied, and the Youngs modulus calculated. An up-to-date version of this instrument has recently been employed by Adam and co-workers [72] for examining synthetic polymer gels near the gel point.

Essentially the same pre-gel information may be obtained by employing a simple "falling sphere in tube" viscometer [73] and this has been used by Richardson and Ross-Murphy for globular protein pre-gels [61]. By timing the fall of the sphere through a fixed distance, the gel time may be estimated as that time when the viscosity η becomes infinite. Of course this type of apparatus has the disadvantage that although low stresses and strain rates are used, the strain γ ($= t \times \dot{\gamma}$) is large and the weak pre-gel network is certain to be disrupted. In the above work, experiments were also carried out in oscillatory strain i.e. γ is small even though ω ($\sim\dot{\gamma}$) was $\cong 1$ Hz. From these complementary approaches different estimates of the gel time t_g could be obtained, reflecting the susceptibility of the pre-gel to breakdown.

Recently, Ellis and Ring [74] have used a ultrasonic pulse shearometer, to study the behaviour of amylose gels (Sect. 4.2.2) at small deformation. Here a square pulse is applied to one piezo-electric crystal, then passes through the sample, and is detected by a second crystal. The modulus of the gel may be calculated from the velocity of the 200 Hz wave. Whilst agreement with static measurements seems fair (10%) 200 Hz, which is equivalent to 1256 rad sec^{-1} for a sine wave, is in the frequency regime where entanglement contributions, and higher frequency intrachain modes, begin to dominate.

By contrast to the above, the falling sphere experiment discussed earlier, exemplifies the strategy of performing experiments at constant stress, and a number of commercial rheometers are now available to measure in this mode. Constant stress transducers are in principle much easier to design than those for constant strain, since in the latter case a large dynamic range is essential. In consequence results from oscillatory stress instruments can be expected to proliferate in forthcoming biophysical literature.

3.2.2 Frequency Dependence

Once the gel cure trace has reached an apparent plateau, the frequency of oscillation may be altered, and G′, G″ charted as functions of radial frequency. This plot is usually known as the "mechanical spectrum". There are two points at issue here, firstly how level is the plateau of the gel cure, and then secondly what does the resultant mechanical spectrum represent.

As far as the curing process is concerned, many biopolymer systems continue to show log G′ increasing as a function of log (t), for days or weeks. This may be due to a number of factors, including loss of solvent, with a consequent increase in the volume fraction of polymer, and ageing. The latter effect, which is particularly noticeable for gelatin, (see Sect. 4.1), is usually attributed to secondary crystallization. However, G′ will eventually tend to decrease with time, because microbiological attack will cause the breakdown of the chain backbone unless small amounts of bactericide are added to the system. For this reason in the (comparatively few) studies which have examined the frequency dependence in detail, measurements have been performed when the trace of Fig. 13 has reached its initial plateau. This at least has the advantage that this time occurs typically after 2–3 hours, making an extensive study worthwhile.

The "mechanical spectrum" obtained after this time has been described by Mitchell as "rather uninteresting" [75]. This merely reflects the insensitivity of such measurements to structural detail. The typical gel spectrum, over the frequency range say 10^{-2} to 10^2 rad s^{-1} consists of two nearly horizontal straight lines [76]. G′ is typically 1–2 orders of magnitude greater than G″, and both may show some slight increase at the higher frequencies. If the frequency range is extended to $> 10^2$ rad s^{-1} (~ 15 Hz), the effect of both "dynamic entanglements" and of more local (and thus more system dependent) intrachain vibrational modes will be seen. Sometimes G″, itself, will show a slight minimum in the experimental frequency range: for example, for gelatin [66]. As concentration is increased the loss tangent, $\tan\delta\ (= G''/G')$ tends to decrease, and eventually becomes zero. This merely reflects the phase angle resolution of most instruments ($\sim 10^{-3}$ rad). Of course for filled gels, or mixed gels, the opposite tendency may be observed, as the presence of "inert material" increases the damping factor. Nevertheless, there is more than an element of truth in the Mitchell criticism, and certainly without other information, the frequency spectrum of a biopolymer gel has only limited value.

The contrast between this type of spectrum, and that found for an entanglement network system is illustrated in Fig. 14; in principle the difference is just that the latter shows flow in the terminal zone region, whilst the former has the "equilibrium modulus" contribution G_e, as originally pointed out for example, by Mooney [77].

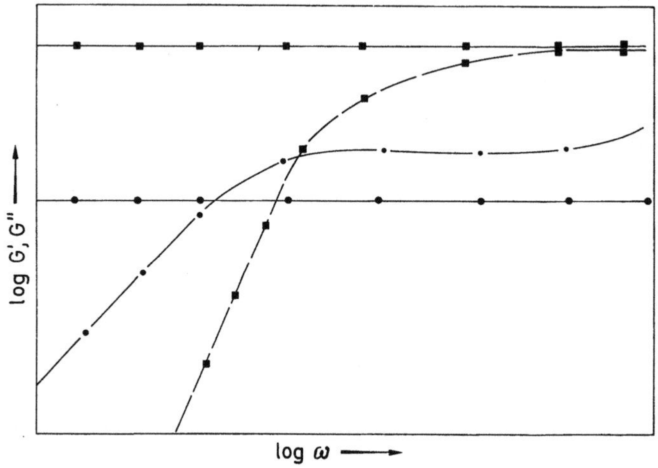

Fig. 14. Mechanical spectra (G′, G″ vs. ω) for a typical gel (solid line), and entanglement network (dashed line), squares G′, circles G″

3.2.3 Strain Dependence

Elastomeric systems show the characteristic stress-strain profile described by the classical theory of rubber elasticity [78]. For these the strain dependence in a small strain oscillatory shear experiment is negligible, and the linear viscoelastic region extends to $\gamma \approx 1$ or greater. Whilst the same independence of G′ is seen for biopolymer gels at $\gamma \ll 1$, the linear region only exceptionally extends to $\gamma \approx 1$. In fact in large deformation tensile experiments (see later), the strain to break normally lies in the range 0–0.5 strain units. A number of the present systems, such as, for example, the extremely heterogeneous coagulates, which can be formed by heating globular protein solutions near the iso-electric pH of the protein, have much more in common with colloidal network systems, (e.g. $\gamma_{\text{linear}} < 10^{-2}$) [79].

3.2.4 Long Time Behaviour

The long time behaviour of biopolymer gels measured either at constant strain, (stress relaxation), or constant stress (creep) has been used by a number of workers to probe regimes of 10^2–10^6 s, in a move to investigate the permanence of physical crosslinks. Recalling the frequency dependence described in Sect. 3.2.2 the creep response in a log-log plot of J(t) against t should be almost horizontal, with an asymptote equal to J_e, the equilibrium compliance Fig. 15. According to Mitchell a rather different picture is seen for a number of polysaccharide gels [80–82], and in creep a terminal flow regime is observed with an apparent Newtonian viscosity $\sim 10^7$ Nm^{-2}. One problem with such measurements is of course that flow and rupture, although different mechanisms, will give apparently the same result. To resolve this ambiguity Ferry [11] suggests that as a rule of thumb, the product σJ_e ($=\gamma_{max}$) should be less than 0.2. In the polysaccharide work it is not clear that this was always the case, so that the measurements reported may have been outside the linear viscoelastic region. A more appropriate route might be to measure the recoverable strain at different stresses. Other workers do, however, support Mitchell's conclusions, although higher apparent viscosities have been reported. The data of Richardson and Ross-Murphy [83] is interesting in that similar effects were observed for some protein gels; for example bovine serum albumin at 10% concentration (w/w) showed terminal flow behaviour, but this decreased very markedly as the effective modulus increased; i.e. the creep compliance became much more ideal as the amount of crosslinking was increased. Moreover for other systems this 'flow' did not occur even after 10^6 s ($\sim$ 10 days). Mitchell himself has pointed out problems in this type of measurement [82], particularly for polysaccharide gels. These include the fact that such gels are very prone to 'weeping' (syneresis). It may be in this case that slippage occurs during measurement, and it is this slippage which is measured rather than creep within the sample bulk. Loss of water is also a problem, although this would decrease the compliance (i.e. increase G(t)) rather than have the opposite effect. Moreover in most of the experiments reported above, the sample has been protected either by using a silicone oil seal, or using a saturated atmosphere. As mentioned before, bacterial attack can also occur over the timescale of days involved in such studies, and this would presumably break down the network in a way reminiscent of oxidative degradation of natural rubbers.

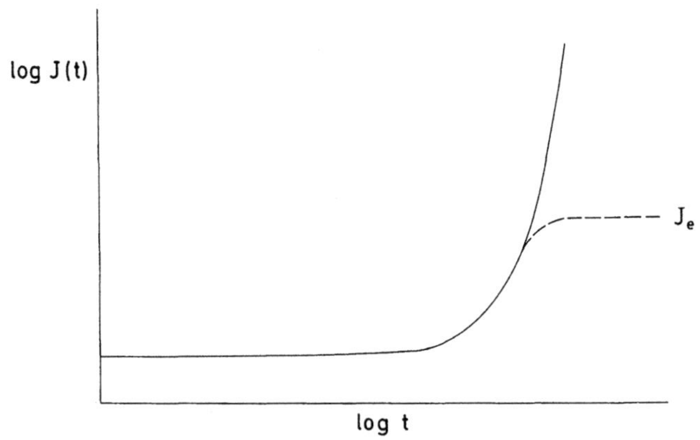

Fig. 15. Creep compliance J(t) plotted against log (t), for a viscoelastic fluid, and for a permanently cross-linked network, with an equilibrium steady state compliance J_e

Overall then, the conclusions about creep and long time flow in such networks must be regarded as equivocal, and each experiment which has demonstrated the phenomenon could be criticised. Indeed, until a completely rigorous regime is adopted, there may be no definitive answer. Apart from sample handling problems there are other experimental difficulties — extremely small strains have to be measured at very low constant stress over long times, and inevitably the long term apparatus stability will be tested *in extremis*. In addition, from a theoretical standpoint, we must be aware that treatments which explain the behaviour of biopolymer gels in terms of permanent crosslinking are likely to be fundamentally at fault. A more realistic model may involve the slow, but noticeable making and breaking of physical crosslinks — in fact some theories including the original approach of Eldridge and Ferry [84] require such an equilibrium to occur. Theoretical consequences of the making and breaking of crosslinks on the overall relaxation spectrum are now considered.

Phenomenological models describing temporary network behaviour occur quite frequently in the literature, from the early works of Lodge, through to those of Curtiss, Bird et al., and Marrucci and coworkers [85]. As far as molecularly (or more strictly macromolecularly) based models are concerned the literature is much less helpful. The concept of reptation of chains above the overlap concentration for linear polymers has, however, been widely applied following the pioneering efforts of de Gennes [86], and of Doi and Edwards [87], and is discussed in recent reviews by Graessley [9,10]. Unfortunately, extensions of this model to branched chains, although quite successful, seem rather artificial in concept [88]. They do, however, predict the very slow diffusion of highly branched chains. For example, the diffusion coefficient of linear chains follows the molecular weight dependence,

$$D \propto M_r^{-2} \tag{5}$$

whilst for branched chains it is found that,

$$D \propto \exp(-M_r)/M_r \tag{6}$$

There are, however, very few models which are able to predict the dynamics, (e.g. the stress relaxation behaviour) of networks when crosslinks are 'temporary' but of non-negligible energy, i.e. they are making and breaking during the time of the stress relaxation experiment. Those that do exist are currently somewhat limited in their application. However in a recent thesis Baxandall [89] has discussed various models, particularly those which include calculations for the contributions of 'added' and 'subtracted' crosslinks (i.e. added and subtracted at different times). Details of the method employed, that of 'replica' elasticity [90], and of the intermediate calculations, are outside the scope of the present article, but for nominally infinite chains, the contribution from removing effective crosslinks modifies the stress relaxation behaviour from the form,

$$\frac{\sigma}{\sigma_t} = \exp(-\mu t) \tag{7}$$

when crosslinks are "added", to a much slower relaxation of the form

$$\frac{\sigma}{\sigma_t} = \frac{1}{(1 + \mu t)} \tag{8}$$

Here μ is a rate constant for the breakage of crosslinks. This apparently anomalous result is a consequence of the memory effect contributed by the removal of crosslinks. Eq. (8) actually gives a relaxation spectrum of the form,

$$H(\tau) = \frac{1}{\mu\tau} \exp(-1/\mu\tau) \tag{9}$$

in contrast to the delta function form for $H(\tau)$ from Eq. (7).
However, Eq. (8) predicts an infinite viscosity, this being a consequence of the original infinite chain assumption. For finite chains, Eq. (9) will have a long time cut-off, so that the terminal viscosity will remain finite. Qualitatively, at least, this form of relaxation seems correct, though contributions from branching and from polydispersity will presumably complicate the quantitative prediction of long term behaviour. Further comparison with the, albeit limited, experimental data available seems timely. Finally, it is interesting to contrast the relaxation implied by Eq. (8) with the early model of Flory based upon a phantom chain model [91]. This gives

$$\frac{\sigma}{\sigma_t} \sim \frac{1}{\mu t \ln(\mu t)} \tag{10}$$

and the terminal relaxation, here illustrated in Fig. 16 is apparently quite similar to that specified by Eq. (8).

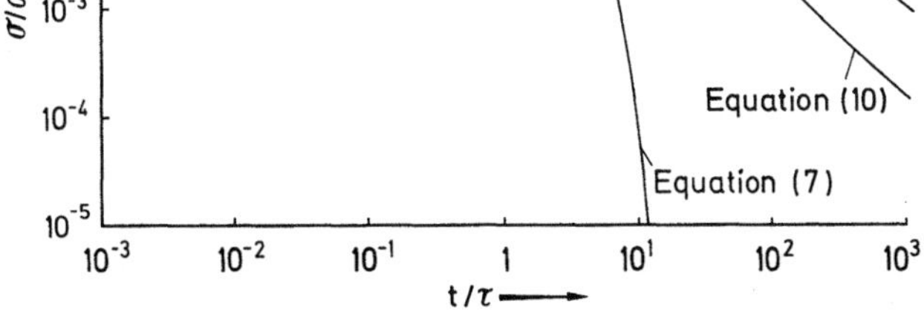

Fig. 16. Relaxation of stress relative to initial stress, plotted against log (t) for Eq. (7), (8) and (10)

3.3 Large Deformation and Failure Behaviour

It is a disappointing, but verifiable assertion that so far relatively little fundamental work has been carried out in this area. One of the problems of course is that whilst for most purposes the only satisfactory experiment involves tensile strain, (this ensures that at the failure region of the sample the strain is homogeneous) the vast majority of data published involve compression or penetration experiments. The reason for this is simple; unless a very sophisticated rig is designed, gel samples can only be tested in tension if they are capable of supporting their own weight, and this is not always experimentally possible. Some years ago, Tschoegl and coworkers [92] overcame this problem by surrounding the test sample (in their case a bread dough) with density matching fluid, so that the effective mass of the sample was reduced. However this procedure has not been widely applied. The usual sample geometry is a dumbbell shape, but defects such as air bubbles can occur, and the need to use an external extensometer to measure the deformation, make the method very unwieldy. Of more serious concern is the limitation that most biopolymer gels are "weak", i.e. at the usual concentrations employed, the mere act of using clamps causes failure to occur preferentially at the clamping point. For semi-quantitative measurements this geometry can be adopted, and the samples may be notched [93]. Clearly little information can then be obtained from the stress-strain profile, but the apparent stress and strain to break may still be computed.

Exactly similar arguments were made by Myers and Wenrick [94] for synthetic elastomer systems in 1974. They compared results obtained with conventional samples with those found using hollow circular rings or ovaloid "race track" shapes, suspended over dowel pins, and then extended. A detailed discussion of the advantages of these latter geometries is included in a series of recent papers from our Group [95,96], and wherever possible we now use the hollow ring configuration. In this way it is very easy to determine stress as a function of strain up to failure.

At the same time, methods developed by those investigating the ultimate properties of gum elastomers, such as the use of the "failure envelope" to present data, have proved adaptable to the biopolymer situation, and these will now be reviewed in some detail.

3.3.1 Stress Strain Behaviour

The statistical theory of rubber elasticity [78] gives the simple form of stress-strain equation

$$\sigma_T = \frac{E}{3}\left(\lambda^2 - \frac{1}{\lambda}\right) \tag{11}$$

with E the Young's modulus, σ_T the tensile stress, T absolute temperature, λ $(= \varepsilon + 1)$ the extension ratio, and ε the tensile strain. This equation implies that the free energy change on straining a polymer chain is due to the restraints placed on configurational rearrangement, and is considered to be totally entropic in origin. At present a number of active approaches are being adopted [97-100] to explain the deviations which occur in real elastomers, but these, although of great interest, are of little relevance to the current area of discussion. All of them have the advantage of treating "minor" deviations from ideality, but in many biopolymer gels, there may be only a very small "entropic" contribution to elasticity, the major effect arising from chains adopting molecular configurations of higher energy, including both conformational energy contributions associated with the deformation of torsion angles, and contributions from the bending and stretching of the semi-rigid structural units. In this case, the deviations from the ideal rubber-like behaviour have to be incorporated either at a purely phenomenological level, or by using conformational energy calculations as discussed by Bailey, Blanshard and Mitchell [101]. In the former case, historically, the most important approach is that due to Mooney and Rivlin [102]

$$\sigma_T = 2(C_1\lambda + C_2)\left(\lambda - \frac{1}{\lambda^2}\right) \quad (12)$$

or other two-parameter equations, such as those due to Blatz, Sharda and Tschoegl (BST) [103] or Kilian [104]. To some extent the adoption of any of these must be of interest only *a posteriori* but the BST equation does allow for gross deviations from Eq. (1) to be included. The BST equation is

$$\sigma_T = \frac{2E}{3n}(\lambda^n - \lambda^{-n/2}) \quad (13)$$

and when it is applied to elastomers, n is always found to be in the range 1.8–2.0. Empirically at least this is not a constraint which need apply to stiff chain networks. An example of the treatment of "entanglement networks" of rods is the work by Doi and Kuzuu [105] who calculated the stress-strain behaviour in such a system. This is discussed in more detail in Sect. 3.7.

3.3.2 The Failure Envelope

One problem in any investigation of ultimate properties is the statistical nature of failure. Even for carefully controlled replicate experiments there will be a distribution of values of e.g. σ_B and ε_B, the stress and strain-to-break. In the classical Griffith's theory of fracture for linear elastic materials, failure is related to the concentration of elastically stored energy at the tip of a macroscopic crack [106,107]. In elastomers failure is said to occur because there will be a distribution of lengths of stress bearing (elastically active) network chains [108]. Each of these has an ultimate extensibility, and as this is exceeded for the shortest chains, the extra stress is "concentrated" on the next shortest chain. In this way the material ruptures by a series of very fast, but apparently random events, just as envisaged in the Griffith's fracture hypothesis.

For the present systems the implication of the above picture should be clear. Since the nature of the network chains in many biopolymer gel systems is such as to furnish only a limited capacity to accept molecular deformation, failure should occur at much smaller absolute strains than for elastomers and the limited data available certainly support this qualitative analysis. Since this result is implicit in the nature of the systems, deviations from linearity in the stress-strain profile will not be very great until the material begins to fail (or shows yield). The recipe for constructing a failure envelope in effect exploits the pronounced strain rate dependence of σ_B and ε_B. If we make

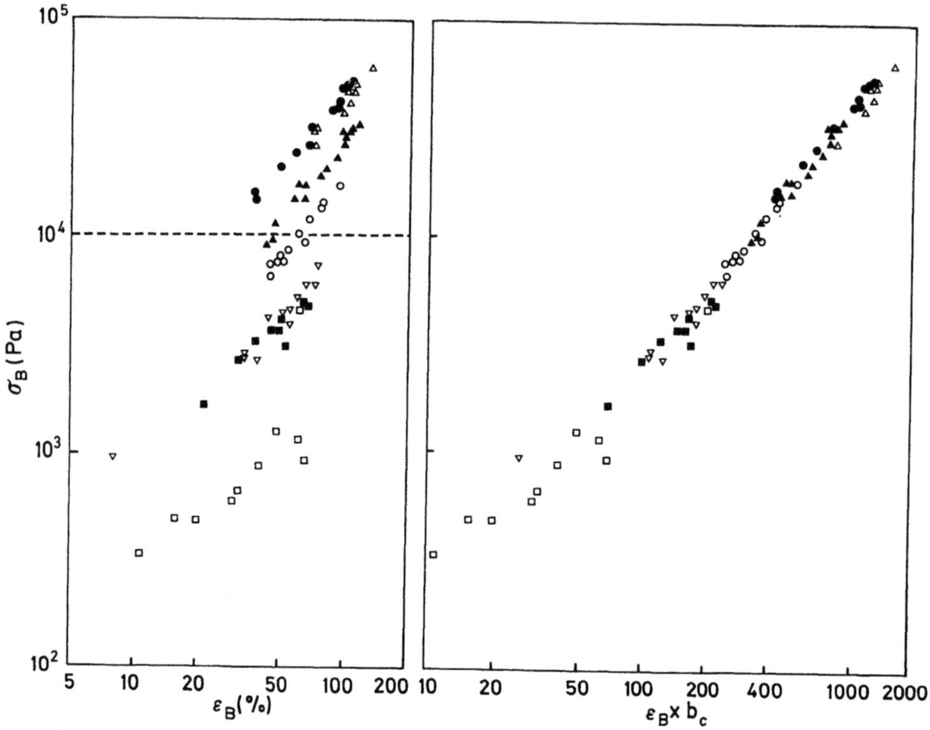

Fig. 17. Failure envelope for gelation gels, concentrations 5%: □, 10%: ▽■, 15%: ○, 20%: ▲, 25%: △●. The dotted line at 10^4 Pa 17(a) [left] is used to illustrate how the data may be superposed by a horizontal shift to give 17(b); (for details see Refs. [95] and [108]. Reproduced with Permission from "Biophysical Methods in Food Research", edited by H. W.-S. Chan, Fig. 21a and b, p. 182–183, Blackwell Scientific, 1984. Copyright 1984 Society of Chemical Industry

measurements of σ_B and ε_B at different extension rates and for different replicates, then we can plot the data collected as in Fig. 17. As was discussed in some detail by Smith and co-workers [108], for elastomers, we can then, in principle, use time-temperature superposition to extend the effective range of extension rates. Qualitatively at least the same manoeuvre may be carried out by using a time-concentration superposition for gel systems and an extended justification for the quantitative adoption of this procedure has recently been published [95]. The success of this strategy can be attributed to two factors. Firstly the separability of the strain and time dependence of $E(t, \gamma)$ is easily shown; for over the experimental timescale the appropriate stress relaxation behaviour — itself a gel characteristic — is clearly seen. Secondly at failure, as discussed previously, the stress-strain profile is not far from linear. Qualitatively, at least, the equivalence of temperature and concentration in such a superposition is, of course, attributable to the free volume.

3.4 The Concentration Dependence of Gel Modulus

The basis for all currently accepted theories of gelation was established in the 1940's by Flory and Stockmayer (FS) [109–112], and a number of papers have appeared recently in which the concentration dependence of the modulus of physically cross-linked biopolymer gels has been discussed from this standpoint. In the random FS f-functional polycondensation model, the critical degree of branching $\alpha_c = 1/(f - 1)$, where f is the functionality (number of functional groups or attachment sites available to form crosslinks) and α is the proportion of functionalities which have reacted. For values of $\alpha > \alpha_c$, a network is formed, whose equilibrium shear modulus may be written

$$G = \Phi N_e(\alpha) \tag{14}$$

where $N_e(\alpha)$ is the number of elastically active network chains per primary chain, and φ is a proportionality constant. For ideal rubbers, φ is given by

$$\varphi = gRT/V_{mol} \tag{15}$$

where V_{mol} is the volume per mole of "primary chain" and g the "front factor", now accepted to be ≈ 1 for entropic networks. For entropic "phantom" networks, g is equal to $(f_e - 2)/f_e$, where f_e is the number average functionality of an elastically active cross-link — see Ref. [1]. The contribution RT represents the entropy gain per mole of network chains on passing from a strained to unstrained state. For the present systems, unfortunately, the major contribution to elasticity is currently assumed to be derived from the adoption of higher energy conformations within strained macro-molecular chains, and the consequent twisting and bending of the overall stiffened chain [113]. Non-ideality in this term has been previously discussed for example for natural rubber, but compared to the present systems, such contributions may be regarded as merely perturbations. A recent detailed Review of thermomechanical contributions to elasticity of networks has been given by Godovsky [113].

The calculation of N_e as a function of crosslinking density may be performed in a number of ways. For example in the Flory treatment [110], if f is large, $\alpha_c \approx 1/f$, and N_e is given by

$$N_e \approx \alpha(2 - W_g)/\alpha_c - 2W_g \tag{16}$$

where W_g is the gel fraction.

For random stepwise polyaddition of an f-functional monomer we can use the extension of the Case-Scanlan network model employed by Dobson and Gordon (cf. Ref. [112]). In these terms N_e is a function of both f and α, and is given by:

$$N_e(f, \alpha) = f\alpha(1 - v)^2 (1 - \beta)/2 \tag{17}$$

where

$$\beta = (f - 1)\,\alpha v/(1 - \alpha + \alpha v) \tag{18}$$

and v the "extinction probability" is given as $v = R(1 - \alpha + \alpha v)^{f-1}$. Here R implies that we take the lowest positive root of ( ).

In connection with the biopolymer gel problem, the model of Peniche-Covas et al. for gelatin [60] is particularly interesting since it allows N_e to be found for an n-chain junction zone, where x such zones occur on each primary chain. In terms of the random f-functional branching model the above treatment gives

$$f = (x - 1)(n - 1) + 1 \tag{19}$$

At this point it is important to re-emphasize that the factorisation of Eq. (14) is not a trivial manoeuvre, rather it is a fundamental step in understanding the nature of the biogel's response to mechanical perturbation. In more detail, we can regard factorisation into the "connectivity" term, i.e. the count of the number of chains which contribute to the modulus, and into the "elasticity", or energy per such chain, to be a general procedure. Such a procedure is not necessary, nor is it normally invoked /for synthetic elastomers, since as already mentioned, these may be regarded as always "ideal" compared to the systems described herein.

To relate the α parameter to concentration we adopt the assumption of Hermans [114], in which the association of primary chains is described by a monomer-dimer equilibrium involving the crosslinking sites (i.e. the functionalities) as first suggested by Eldridge and Ferry [84] — see Section 3.5. This allows us to write the dissociation equilibrium constant for a link $\rightleftharpoons$ two free sites in an Ostwald dilution law form viz.

$$K' = [(1 - \alpha)^2/\alpha]\, N_0 f \tag{20}$$

Here N_0 is the number of primary chains per unit volume, and f is the functionality. For the case of "weak binding", i.e. K' presumably $\gg 1$, then clearly $\alpha \ll 1$, and this can be solved to give

$$\alpha = \frac{fN_0}{K'} = \frac{fKC}{M_r} \tag{21}$$

as was first carried out by J. Hermans in 1965. In Eq. (21) K is the association constant for cross-linking ($1/K'$) and M_r the primary chain molecular weight and from this it is easy to see that

$$\frac{\alpha}{\alpha_c} = \frac{C}{C_0} \tag{22}$$

Since $N_0 f$ is also the number of sites per unit volume, we can say that $\alpha \propto \varphi$, the volume fraction of reacted sites, and thus

$$\frac{\alpha}{\alpha_c} = \frac{\varphi}{\varphi_c} \tag{23}$$

as deduced by a number of workers [62, 115].

From the above equations is clear that for large f, $C_0 = M_r/Kf^2$ and for high functionality, Hermans assumption is thus implicitly the above equivalence of α/α_c and C/C_0. C_0 is the critical concentration below which no macroscopic gel is formed under the prevailing experimental conditions. We reiterate that C_0 is *not* the same as the overlap concentration C* given by $C^* \cong S[\eta]$, where S is a constant ~ 1. In practice, we expect $C_0 < C^*$, and this is easily demonstrated for globular proteins ($[\eta] \sim 0.03$), whence $C^* \cong 33\%$ w/w, whilst $C_0 \sim 1 - 10\%$ w/w (depending upon conditions).

The existence of such a critical concentration in an aggregating biopolymer system is important in practice, and the equilibrium assumption predicts this property quite naturally. In the case of synthetic polymer gel formation where covalent bonds usually form irreversibly, the existence of a critical concentration is attributable to competition between intra and intermolecular reaction at low concentrations [116]; of course some intramolecular crosslinking will also take place in the present systems, but a rescaling relative to the critical gel point conversion should reduce this effect [117].

If, rather than applying the equilibrium assumption to the "high functionality" case, we use Eq. (20), and make a full solution for α valid for any f and K, we can derive the more general results

$$C_0 = M_r(f - 1)/Kf(f - 2)^2 \tag{24}$$

and $\alpha/\alpha_c \neq C/C_0$.

We now replace the factor, g, in Eq. (15) by "a", a generalised front factor, to reflect major deviations from ideal rubber elasticity and assert that in general a > g. Whilst there is currently no way of calculating "a" independently, recent work by Mark and Curro has discussed the stress-strain isotherm at finite strain, and how it is affected by the presence of short (i.e. non-Gaussian) network chains [118,119]. Deviations from ideal behaviour occur much more pronouncedly for these short chains, as confirmed by both Monte Carlo and model network studies, and this lends support to our assertion above. The same effect may be deduced from the calculations for stiff polysaccharide chains discussed earlier [101].

Using Eqs. (15) and (17) and collecting all terms together we have a final expression for the relationship between G and C as a dimensionless function of the form [70].

$$\frac{GK(f - 1)(f - 2)}{aRT} = \frac{C}{C_0}(f - 1)^2 \alpha(1 - v)^2 (1 - \beta)/[2(f - 2)] \tag{25}$$

For any choice of f, experimental G vs. concentration data may be reduced to master curve form by optimising "a" and C_0. The changing shape of the cascade master curve as f varies is indicated in Fig. 18. Hermans' master curve is similar, but not identical to, the high f cascade limit, and in both cases a limiting C^2 concentration dependence is predicted, although for the lower functionality cascade models, the exponent is substantially less than 2.0, (i.e. ~ 1.1 for $f = 3$). Of course in any case for high C/C_0, network defects, entanglements etc. will become increasingly significant, so that limiting behaviour should not be regarded as a fair test of the treatment.

It is interesting to note that Johnston [120] has used essentially the Flory vulcanisation model with $a = 1$ to model the kinetics of shear modulus development of casein-rennet gels (Sect. 5.4.2). He also reports the $G \propto C^2$ dependence at high concentration.

At this point it is appropriate to discuss the historical significance of this exponent; Hirai suggested many years ago a dependence $G \sim C^{\delta}$ [121] with $\delta \sim 2$. This limit is reached for the high functional models for $C/C_0 \sim 10$; higher exponents are sometimes reported, but clearly only represent the limited range of the data collected. More recently a limiting behaviour $G \sim C^{9/4}$ has been suggested [41] by noting the parallel between the osmotic pressure of semi-dilute polymer solutions and the modulus of gels, the "osmotic scaling" relationship. This relationship does, of course, fit limiting data as well as (and sometimes better) than the $G \sim C^2$ relationship, but if the contribution from e.g. the trapped entanglements [122] is also included in the C^2 dependence then a slightly greater exponent than 2 would in any case be expected.

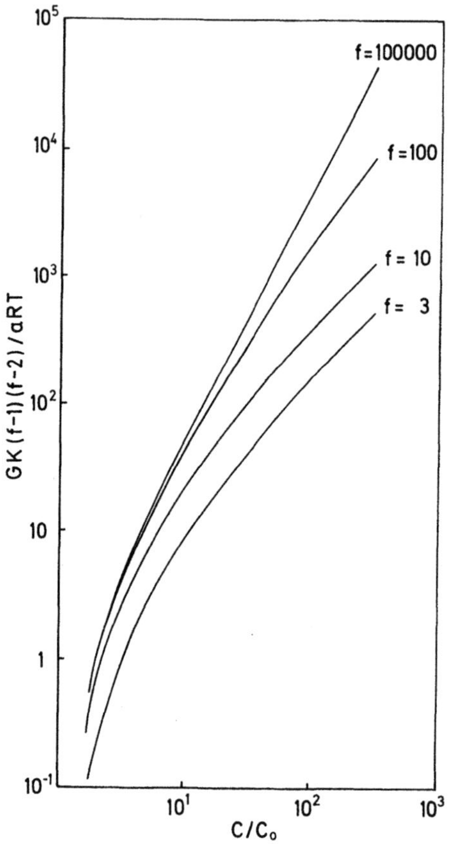

Fig. 18. Reduced modulus (Eq. 25) plotted against log (C/C_0) for different functionality, f. Herman's function is similar, but not identical to the high f cascade limit. Reproduced with Permission from British Polymer Journal *17*, 164 (1985), Fig. 1, p. 165

Very recently Oakenfull [123,124] has discussed a slightly different model which incorporates most of the features above, including the equilibrium assumption for the making and breaking of junction zones, and the length and number of these junction zones. This has been applied to data for high methoxy pectin gels and to gelatin itself [123]. In many respects the method is very similar to both the generalised Hermans method [70,114], and to the model of Peniche-Covas et al. [60], when the various parameters are re-interpreted. Oakenfull uses the expression

$$G = cRT/M_c ,$$

with

$$M_c = f(M_r, M_j, K_j, n) \tag{26}$$

where M_c is the molecular weight between crosslinks as discussed by Flory and coworkers [109] (and in the terms used here $M_c = f(\alpha)$), M_r is again the primary chain molecular weight, M_j is the molecular weight of the junction zone region, K_j is an equilibrium constant for the formation and dissolution of junction zones, and n is the "molecularity" of the helical junction zones (i.e. 3 for gelatin, 2 for pectin), just as in the Peniche-Covas model. Re-interpreted in terms of the latter model x, the number of such zones is just M_r/M_j.

However, although using the same equilibrium assumption as was adopted previously, Oakenfull has extended the treatment from the situation where equilibrium involves pairs of crosslinking sites $\rightleftharpoons$ crosslinks, only, to one where n can be 3 or more. By allowing $n > 2$, as suggested above, an element of cooperativity is included. K_j then is given by

$$K_j = [J]\, M_j^n/[n(C - M_j[J])]^n \tag{27a}$$

with

$$[J] = K_j[L]^n \tag{27b}$$

Here [J] is the concentration of junction zones, and [L] the concentration of crosslinking loci. In these terms the critical concentration C_0 is equal to $M_r[J]$, and [J] is found from Eq. (27a). The limitations of the Oakenfull treatment must be seen from the front factor assumption $a = 1$, although the general assertion that $a \neq 1$ could clearly be included. In practice the method shows the same qualitative behaviour as the other approaches described here, with $G \sim C^\delta$, $\delta \to \infty$ as $C \to C_0$ from above, Fig. 19. Interestingly for large C/C_0 the asymptotic behaviour at high functionality $\delta \to (2n - 1)/n$ is predicted (i.e. δ is always <2).

The problem, as with the other approaches, is that generality can only be obtained

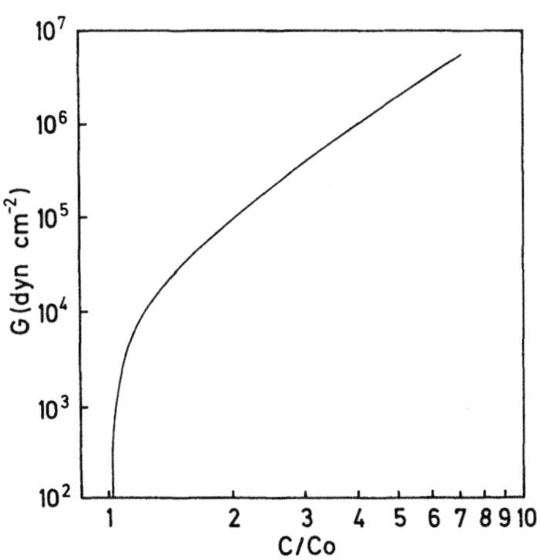

Fig. 19. Modulus plotted against $\log(C/C_0)$ according to the model of Oakenfull (Ref. [124]). Here the parameters n, K_j and M_j are 3, 1000 and 1000 respectively

by increasing the number of unknown parameters, when quantitative understanding and future progress really requires that the number of parameters should be reduced. For this, measurement on a range of fractionated samples of low initial polydispersity, and known M_r seems to be essential.

Recently Tokita, Niki and Hikichi [62] have made very careful measurements on casein gels (see Sect. 5.4.2) very close to the critical concentration using an ultra sensitive oscillatory Couette rheometer. In this way they were able to very accurately measure G* at low strain, with moduli $<10^2$ dyne cm^{-2}. They fitted the data with the percolation model (for details see Ref. [125]). This is in essence similar to the random polycondensation treatment, but whereas the simple form of the latter neglects the effects of intramolecular crosslinking, the percolation model includes these effects, although not in a realistic way for real systems, since it normally involves the connection of edges or vertices on a rigid lattice model [125]. Almost all the published work on percolation models addresses only the question of exact values of critical exponents, i.e. the value of m in $G \propto (\alpha/\alpha_C - 1)^m$ as $\alpha/\alpha_c \to 1$.

However, it has been shown [126] that very close to the gel point, for dilute systems, and where intramolecular reaction is most encouraged, the percolation model fits better than that of Flory and Stockmayer. For more concentrated systems the latter is more appropriate and for the stiffened "chains" discussed here, the cyclisation effect will in any case be reduced. The main limitation of the percolation theory, in this context, is that firstly it generates results only extremely close to the gel point, and secondly it can make no estimate at all of the proportionality constant in Eq. (14), and treats it as an adjustable parameter. Although in principle different models for gelation, such as those discussed so far, may be distinguished using rheological data, this is currently an ideal situation, both uncertainty in the measurements, and in the nature of the sample, making such distinctions unreliable in the majority of cases.

Indeed using the Tokita data, Gordon [115] has demonstrated how a variety of different models, e.g. aggregation of casein molecules, or aggregation of preformed casein micelles, can be applied to the same gelling system. At the same time the gelation may be treated as either a random polycondensation, or a vulcanisation of preformed linear chains. In practice Gordon found that all of these models, including the percolation model fitted the data just as effectively. Clearly, in the casein aggregation case more structural information is required to distinguish between the models, and this is discussed more fully in Sect. 5.4.2.

In Sect. 3.3, we discussed a purely phenomenological description of the large-deformation behaviour of biopolymer networks. In later Sections we shall describe in more detail its application to, for example, agarose and gelatin gels. Nevertheless a brief qualitative discussion of the molecular aspects of large deformation behaviour is appropriate here, particularly in relation to concentration effects. If all the strain energy applied, for example, in uniaxial tension, is stored in the elastically active chains, then since we assume that the modulus is proportional to the density of these chains, normalisation of the stress by E, the Youngs modulus, should give the same functional form for the given system, regardless of concentration [95]. By contrast, if other contributions to the strain energy are important, they might be expected to manifest themselves as a concentration dependence of the strain function. For gelatin samples at least, no such obvious effect is seen, presumably because over the measured range (5%–30% w/w) they are all substantially solvated gels.

However, the break stress, appears to depend upon strain rate ε in an unexpected way, often showing a maximum at the lowest ε. In a recent paper [95] we have attributed this to an increase in helix-helix, or "quasi-crystalline", intermolecular associations between chains induced by strain orientation. If these associations are themselves relatively weak, the relief of stress locally by dissociation will require less energy than further chain extension, increasing the effective local extensibility — "unzipping" — by increasing the length of disordered chains. The application of X-ray [127] or NMR [128] techniques to strained networks, as used in studies of synthetic elastomers, could prove of value here.

At this point it is perhaps appropriate to discuss the preceding Section in terms of the recent progress in the theory of rubber networks. Although this latter topic is not of major importance to the present text, it is important to reconfirm that we have adopted a fairly primitive "lower bound" to the current thinking on rubber networks. We feel that currently this is justified by the dearth of rigorous experimental studies, but in the future hopefully progress on this front will be forthcoming.

To illustrate this point in more detail we could re-examine Eq. (14). This assumes that for covalent networks the shear modulus is proportional *only* to the number of elastically effective (covalent) cross-links. Some years ago it was suggested that an extra contribution to the modulus should accrue from those physical entanglements made permanent by subsequent crosslinking along the chain contour. These are known as "trapped entanglements", and their probability of occurrence was first calculated rigorously by Langley [122]. What remains equivocal is whether or not such a contribution contributes to the equilibrium modulus of the network — this remains an area of controversy, and experiments on the same system e.g. by Macosko and by Mark and co-workers come to no definitive conclusion, (cf. Ref. [1]).

At the same time the fundamental "chemical" model of rubber networks has been challenged by workers using a constrained "tube" approach, analogous to that originally adopted for entangled linear polymers. In these tube models one single chain is said to be surrounded by an effective medium (the other chains in the system) which resists the motion of the original chain except at its ends. This slow motion of the original chain along its own length is known as "reptation" (such models are discussed in detail in Ref. [2]). The argument follows that since the tube model can give a good description of viscoelasticity it should be capable of calculating elasticity itself in systems with covalent crosslinks. A number of papers have appeared using tube models (although of subtly different form) to calculate rubber elasticity, and this area is one of high current interest. The applicability of all the above approaches to networks in which the original crosslinks are not permanent, but are nevertheless long-lived is not, however, currently clear. Aspects of this model are discussed in Sect. 3.2.4.

3.5 The Temperature Dependence of Gel Modulus

No review of the properties of biopolymer gels and networks would be complete without a mention of the classic work of Eldridge and Ferry [84,129] on the dependence of the melting point of gelatin gels on concentration and on primary chain molecular weight. They established that a plot of ln (C) versus $1/T_{mK}$ (T_{mK} = melting temperature in degrees Kelvin) was almost linear, as was ln (M_r) versus $1/T_{mK}$. This suggested that a van't Hoff law was in evidence, which could be written approximately as

$$-\left[\frac{d \ln C}{dT_{mK}}\right]_{M_r} = \frac{\Delta H^0}{RT^2} \quad (28)$$

where dT_{mK} was the change in gel point "temperature". Since the gel point itself was assumed to occur at constant degree of crosslinking (i.e. no intramolecular "wastage" of crosslinks), this immediately suggested that the formation of the gel was governed by an equilibrium between

2 crosslinking sites $\rightleftharpoons$ 1 crosslink

ΔH^0 being the enthalpy for this process. Integrating Eq. (28) gives the observed gel melt temperature/concentration dependence. This equilibrium assumption is widely used in the description of thermoreversible gels (see last Section on concentration dependence of the modulus), although the reality of such an equilibrium is not well established. However the implied dependence of the density of crosslinking on temperature may be used to rationalize the observed temperature dependence of mechanical properties.

For an ideal ("entropic") rubber (just as with the pressure of an ideal gas), $G \propto T$. In practice, even for elastomers, this behaviour is not often observed, since, for example, for natural rubber increasing the temperature decreases the fraction of crystalline material. The net result is that over narrow regimes of temperature, the ideal behaviour is rarely seen, except for model elastomer systems, e.g. siloxanes [130]. For thermoreversible gels, the decrease in the number of crosslinks with temperature is usually much more pronounced than any increase from the above entropic contribution, and this effect has been used to rationalize the common experimental observation that G decreases with increasing temperature. Of course, as outlined earlier, the assumption of a purely entropic contribution to the elasticity of a network chain is almost certainly invalid for most biopolymer gels. Unfortunately this assertion cannot really be tested for a thermoreversible network, since as suggested above the number of crosslinks will also be decreasing strongly with temperature. For some non-melting heat-set gels, however, such effects can be examined in more detail. For example by cycling the temperature until no further increase in G was observed at a fixed higher temperature, a study of globular protein gels gave an apparent $G \propto T^{-5}$ dependence (over the range 35°–85°) [83]. Discussion of these effects has focussed on the "enthalpic" contribution to elastic energy, which would have a negative temperature coefficient, as observed. Similar effects are expected for non-covalently crosslinked thermoreversible networks of synthetic polymers, for example PVA and PVC gels [131, 132].

For the thermoreversible gel case, Nishinari et al. [133] have formulated a new approach to the modulus-temperature relationship by assuming that a gel consists of Langevin chains (to allow for finite extensibility), with both ends bound into a junction zone region. The ends are assumed to be progressively released from these regions with increasing temperature, by "sublimation". In this way they were able to calculate the temperature dependence of E (the static Youngs modulus) for such gels, and compare results with their own data for PVA networks. This temperature dependence is a very sensitive function of a constraint release parameter μ', the ratio of the number of chains released from the junction zone to the RMS end-to-end distance. Over a limited range, E appears to be a parabolic function of T, at first increasing and then decreasing again.

No discussion of the temperature dependence of the properties of thermoreversible gels could omit mention of the helix-coil transition, since in a number of cases the value of T_{mK}° (i.e. the hypothetical melting temperature of the gel at zero concentration) corresponds to that close to, or at the midpoint of the helix-coil melting curve, that is the mid-point temperature, T_m. Particularly for gelatin and for the polysaccharides agarose and the carrageenans, the primary stage in gelation is believed to involve multiple helix formation. Since this involves the adoption of specific conformations, in which asymmetric residues are involved, there is usually a marked and discontinuous change in observed optical rotation when charted as a function of temperature (see Sect. 2.2.1). The sharpness of this transition reflects the cooperativity of the helix-coil transformation [134].

The theory of the helix-coil transition for biopolymers has been extensively developed over the last quarter of a century, and has frequently been discussed in the literature [135], so that coverage here will be brief. The simplest case is for a single helical chain $\rightleftharpoons$ coil equilibrium, and theory relates the fraction of residues in the ordered state in any situation to an order parameter (s). This fraction f' is given approximately by

$$f' = \frac{1}{2} + \frac{(s-1)}{2[(1-s)^2 + 4\sigma s]^{1/2}} \tag{29}$$

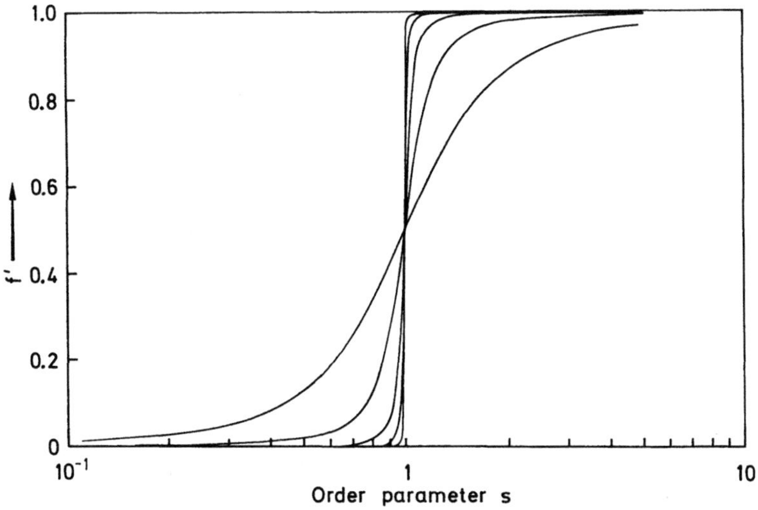

Fig. 20. f', fraction of helical segments, plotted against order parameter, s for values of the interruption constant σ of .1, .01, ... , 10^{-5}, using Eq. (29). As σ becomes smaller the transition becomes more and more discontinuous

where σ is the interruption constant (here assumed to be independent of the length of the interruption). For infinite chains, the helix-coil transition in this case reduces exactly to the one-dimensional Ising model description for ferromagnets, and Fig. 20 illustrates the dependence of f upon s for different values of the σ parameter. Since $\ln s \propto 1/T$ (van't Hoff), for a limited range of T, $s \sim 1/T$ i.e. $\ln s \sim -T$, and thus this plot mirrors the temperature dependence of the helix-coil transition. For more realistic, but more complex, models such as those involving multiple helix formation, when the chains can be either "matched" or "unmatched", or when the chains are finite in length, some of the features of Fig. 20 are preserved even though the symmetry about $s = 1$ is no longer obtained. Such refinements have been described in detail e.g., in Ref. [135].

For agarose (Sect. 4.2.1) the hysteresis between the gel "setting" and gel "melting" temperatures is particularly pronounced, since typically a gel formed at ~30 °C will not melt below ~80 °C [136,137]. Since the junction zones of this gel are believed to involve double helices, the ordering process itself will be concentration dependent, but this large hysteresis is attributed to the further aggregation of the ordered regions. Similar, but not such dramatic effects are seen in data for carrageenan gels.

Both the contribution from the ordering of sections of chains, and from the subsequent aggregation of these regions, are counted in the α parameter of the Flory-Stockmayer theory (Sect. 3.4), but currently there is no way to separate these quantitatively. However, the dependence of G upon temperature may be sketched in the form below

$$G \propto (\alpha/\alpha_c - 1)^n \qquad [n \geqq 2] \tag{30a}$$

(α_c = the critical gel conversion, and $\alpha/\alpha_c > 1$).

$$\alpha \propto s\ (+\text{ contribution from aggregation}) \tag{30b}$$

with s the order parameter and

$$\ln(s) \propto \frac{1}{T}\ (\text{as above}) \tag{30c}$$

The total contribution (30b) must include the Eldridge-Ferry "melting" terms already discussed above. However from this very simple argument, in a region not far from (but not *at*) the transition temperature, $G \propto (1/T)^\varepsilon$, since the Eldridge-Ferry contribution will also produce a van't Hoff temperature dependence.

3.6 Phase Behaviour and Mechanical Properties

Clearly in any discussion of the behaviour of highly solvated gels (typically $\varphi_s \gtrsim 0.9$, where φ_s is the volume fraction of solvent) we should not ignore the presence of this solvent or its interactions with the biopolymer network. The equilibrium swelling of networks is normally discussed by equating the polymer solvent mixing terms (making the system swell) to the constraints applied by the network itself. For synthetic elastomers in non-polar solvents this treatment is quantitatively successful. Unfortunately, however, there are practically no data for the thermodynamics of mixing of biopolymers with aqueous electrolytes. The data which do exist seem so contradictory that, for example, the Flory-Huggins χ parameters for almost all the systems discussed in this article are effectively unknown. Despite this obvious practical limitation, the approach to the case of gelation with concomitant phase separation in thermoreversible networks has been discussed theoretically by a number of workers. For example, following Dušek [138] it is possible to use a mean field approach of the Guggenheim-Tompa type, whereas Coniglio, Stanley and Klein [139] have used a percolation model.

In the mean field theory, analogy may be made to the theory of regular solutions and to the "quasi-chemical" equilibrium approach of Guggenheim. In this we assume two species A and B (so far unspecified) can "react" in the form

$$AA + BB \rightleftharpoons 2AB$$

The breaking of an AA and BB pair results in an increase in the internal energy of the assembly by an amount given by

$$4X_{AA}X_{BB} = (X_{AB})^2 \exp(-2W/kT) \tag{31a}$$

where X_{AA} is the mean number of AA pairs, and W is the non-bonded energy of the AB pair. For example A could be solvent species and B a polymeric species. In the latter case the procedure of Tompa for the lattice model for molecules occupying more than singly connected lattice sites may be adopted, and thus the thermodynamics of both linear, branched and cross-linked solutions may be treated. For the case of gel and solvent molecules, this theory of Dušek [138] is exactly the mean field analogue of the Coniglio-Stanley-Klein theory discussed below.

Coniglio, Stanley and Klein (CSK) [139] have discussed a lattice model for solvated networks which involves the correlated binding of solvent with the solute, and which combines the connectivity and thermodynamic terms. In more detail a lattice graph is constructed, and both solvent and polymer (network) molecules occupy sites on this lattice, but the "original monomers" and solvent molecules are not distributed at random on these sites. Instead a correlation of position is assumed. The "monomers" are allowed to interact by both non-bonded van der Waals and the bonded interaction of covalent links. Then assuming only nearest neighbour interactions on the lattice it is possible to write

$$p_B = \frac{(1-\xi)\exp(-E/kT)}{\xi \exp(-W/kT) + (1-\xi)\exp(-E/kT)} \tag{31b}$$

where E is the bonded energy (weight $1 - \xi$) and W the non-bonded energy (weight ξ) — the model is illustrated in Fig. 21. Here p_B is the probability that a particular bond is present between two nearest-neighbour monomers. This model strictly applies only for a system where bonds are being created and destroyed on a time scale short compared to those of most observations. The CSK model then allows, from the temperature dependence of p_B, the phase diagram to be calculated as a function of both the temperature and "connectivity" dependence. The predicted dependence of this

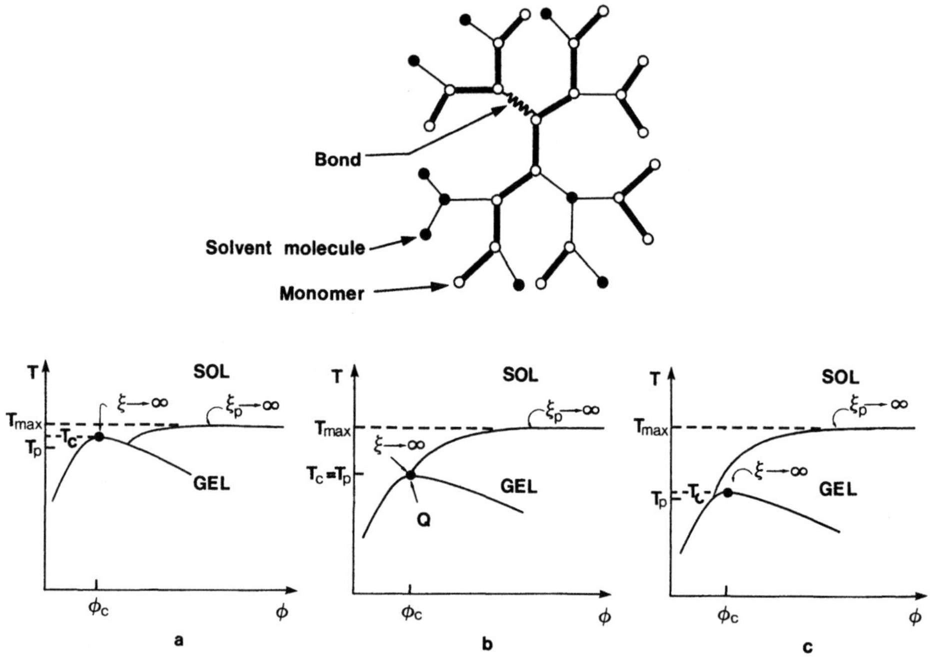

Fig. 21 a–c. Site bonded correlation model of Coniglio, Stanley and Klein (top) and phase diagrams predicted, as a function of volume fraction of polymer, φ. ζ is the correlation length, and ζ_p the connectivity length. The diagrams correspond to different ratios of the critical temperature, T_c to the gel melting temperature, T_m. Reproduced with Permission from Physical Review B, Figs. 14 and 15, p. 6816 and 6817, 1982

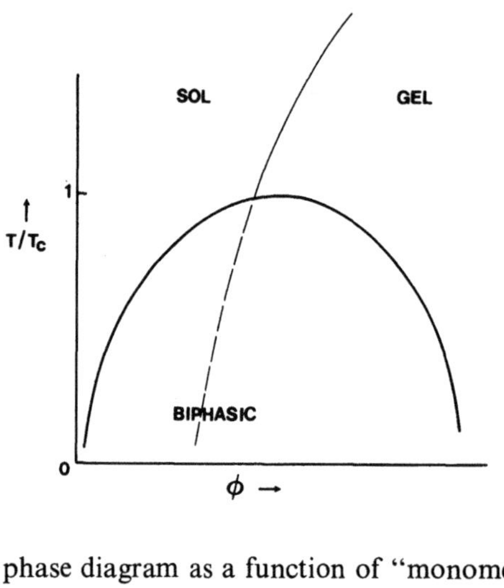

Fig.–22. Simplified phase diagram from the CSK model (cf. Fig. 21); the sol/gel line is extrapolated (————) into the two phase region

phase diagram as a function of "monomer density" d is illustrated in Fig. 22 (Such a system is that designated by de Gennes as a "weak gel" [41]). Note that in this review we use a different and more specific definition of "strong" and "weak" gels which distinguishes, using the operational definition of a gel from mechanical spectroscopy "finite energy" networks from those which are transient in time). From Fig. 22 it may be seen that the sol/gel curve is qualitatively the same as that of Fig. 18. The CSK treatment is noteworthy in the context of thermoreversible biopolymer gels since it would appear to be the first attempt to combine thermodynamic and connectivity effects for those systems and it does allow the above phase diagram to be constructed for a gel as a function of temperature. Tanaka et al. [140] have applied the CSK method to data for gelatin gels in alcohol/water mixtures.

As outlined earlier, treatment of the thermodynamics of thermoreversible gels has been based upon the Eldridge-Ferry theory in which the "melting" temperature, i.e. the sol/gel transition temperature, of a thermoreversible gel is treated in an analogous way to the melting of a partially crystalline polymer, as described by Flory [109,141]. This relates the melting temperature T_{mK} to concentration (volume fraction of polymer) by the following relationship.

$$\ln (C) = \Delta H_m / RT_{mK} + \text{constant} \qquad (32)$$

where ΔH_m is the dissociation enthalpy of junction points. Takahishi et al. [142] have combined the Eldridge-Ferry gel melting theory, with the Flory crystalline melting point equation to derive a result for a "fringed micelle" network of amorphous and crystalline regions. This is given by

$$\frac{1}{T_m^g} - \frac{1}{T_m^c} = R \ln (x_i)/(\Delta H_u - 2\sigma_e) \qquad (33)$$

where ΔH_u is the enthalpy of melting, σ_e is the end group energy ($\Delta H_u - 2\sigma_e > 0$) and T_m^c and T_m^g are the melting temperatures of the purely "crystalline" system and of the gel (mole fraction x_i) respectively. This equation has been very successfully applied to a number of gel systems and a very similar formulation has been used by Donovan [143] to describe the gelatinization temperature of starches. Again this increases with concentration (i.e. with the concentration of "crystallites"), in a way consistent with the Flory melting point equation.

3.7 Properties of Concentrated Rod Systems

Since a number of the biopolymer network systems discussed later are formed from stiff or worm-like chains, often generated by coil-helix transitions, a brief digression into the expected properties of a concentrated system of rods is appropriate. This is an area of extremely rapid progress, spurred by the great interest in polymer liquid crystals [144], and in viscoelastic models for anisotropic chain motion in more concentrated solutions, such as those due to Doi and Edwards for rod and coil polymers [87,105,145].

The theory of solutions containing rigid rods was due originally to Onsager. The extension by Flory [146] predicts that an isotropic-anisotropic phase transition occurs (above some critical concentration (volume fraction) φ_p^*, given by $\varphi_p^* = (8/p)/(1 - 2/p)$. Here p is the axial ratio (L/d) of the rod. For $\varphi_p \approx \varphi_p^*$, the system spontaneously separates into two phases, the "dilute" phase remaining isotropic, whilst the "concentrated" phase consists of an anisotropic arrangement of aligned rods. The transition from disorder to order is driven by the competition between the "orientational entropy" (minimized by an isotropic distribution of rods) and their "translational entropy" (minimized by a parallel array of rods). If φ_p is increased further, the phase volume of anisotropic phase increases until it becomes unity. The relationship above has been extended by Flory in collaboration with Warner [147] to allow for some chain flexibility (which tends to increase φ_p^*) whilst an alternative treatment by Khoklov and Semenov [148] based upon a wormlike chain model gives slightly higher values of φ_p^* than these approaches. In practice with model rod polymers, only approximate agreement with theory has been found, but this most probably reflects the difficulty in preparing very narrow dispersity samples, the tendency to form aggregates, and also the susceptibility to shear-induced anisotropy. This last effect is particularly marked for isotropic solutions close to φ_p^*.

Doi and Kuzuu [105] have calculated the stress-strain profile for isotropic glasses based on rod-like elements with a volume fraction φ_p given by $1/p < \varphi_p \ll 1$. In this model there are no cross-links as such, because two contacting rods can slip almost freely at the contact point, but nevertheless there will be a macroscopic elastic response resulting from the energetics of bending of the rods themselves. This model produces a stress-strain profile which is S-shaped both in uniaxial tension and in shear (i.e. opposite to that seen for rubbers). The intermediate stress "hardening" regime occurs because the larger the deformation the greater the number of contact points between the rods, and thus the greater the stress needed for further deformation. The final "softening" reflects the saturation of the number of contact points. The initial modulus of the system is also extremely small. According to Doi and Kuzuu the S-shaped profile is often found in gels of biopolymers. However they cite only data for the chromatin fibre studied by Vorob'ev [149], which is an extremely specialized system. In our opinion such behaviour is not common for gels, in fact the more usual profile is one of monotonically decreasing slope up to failure. Nevertheless one system where this sigmoidal profile does appear to occur is in the fibrin-fibrinogen networks (see Sect. 5.2.5). Here the fibrinogen macromolecule is known to be rod like, and for a number of years fibrin gels were belived to be formed purely by "entanglement" exactly as described by the Doi-Kuzuu model.
Recently, however, light scattering and stereo electron microscopy by Müller and workers seem to have positively demonstrated a branching process, and subsequent lateral aggregation, i.e. that the structure is that of a true crosslinked network, (cf. Sect. 5.2.5).

A further class of materials which have been reported are the so-called 'Miller' gels formed, for example, by the aggregation of helical rod-like polypeptides. The details of these systems are described in Sect. 5.5.1, and it is interesting that Miller and coworkers explain their formation as a consequence of a spinodal decomposition mechanism. The spinodal is defined as the limit of metastability, i.e. $\partial^2G/\partial C^2 < 0$, where $\partial^2G/\partial C^2$ represents the second derivative of free energy with respect to concentration. In the Cahn-Hilliard [150] mechanism, phase separation occurs without nucleation, and produces a periodic structure with a characteristic spacing of the spontaneous "concentration waves" which develops with time. If one of the phases is sufficiently rigid, or sufficiently concentrated, to be "glass like", the periodicity is frozen into the system, and is reflected in the characteristic 'spacing' of the network.

These systems are also unusual in that they show very little concentration dependence of the modulus, so that for PBLG (polybenzyl-L-glutamate), $G \propto C^{0.2}$. The rational explanation for this would appear to be that very few extra crosslinks are formed at higher concentrations because the system is already glasslike under the kinetic control. Instead, existing 'fibrils' become thickened. In terms of our original discussion, the small subsequent increase in modulus is due only to the increase in the parameter 'a' and not to the increased connectivity of network chains.

4 Networks from Disordered Biopolymers

4.1 Gelatin Gels

4.1.1 Introduction

Gelatin [151], a proteinaceous material derived from naturally occurring collagen (see Sect. 5.5.3) is a substance of great commercial importance. The gelling properties of hot solutions of gelatin, when cooled, are well known, and are widely made use of both in industry and in the home.

Gelatin derives in particular, from the fundamental molecular unit of collagen, the tropocollagen rod [151, 152]. This is a triple helical structure (Fig. 23) composed

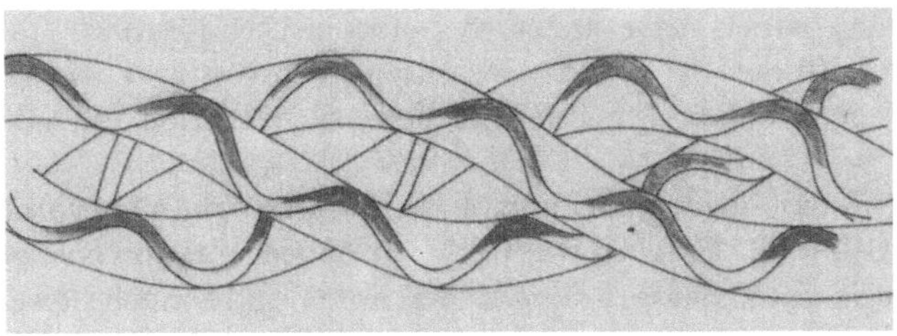

Fig.-23. Schematic representation of triple helical peptide chain structure of the tropocollagen rod

of three separate polypeptide chains, the molecular weights of these being ~110,000 daltons. Although the exact amino acid content of these chains, and sequences, vary from one type of collagen to another, they always contain large amounts of proline and hydroxyproline as well as large amounts of glycine. Other residues also present include polar amino acids which tend to occur in positive and negative clusters at points along the tropocollagen rod surface. The proline content of tropocollagen is particularly important, as it tends to promote special helicity in the individual polypeptide strands (polyproline II helix) and ultimately, the unique triple helical structure of the tropocollagen rod. The proline distribution, which achieves this, is believed to be non-random, with proline and hydroxyproline residues occurring together.

Both commercially, and in the laboratory, gelatin is derived from collagen by hydrolytic degradation [151]. In industrial processes both acidic and alkaline treatments are used, and raw collagen comes from a variety of sources, for example, from hides and from bone. Industrial gelatins are complex, having a whole spectrum of molecular weight components, including those based on single peptide chains (α-gelatin), two chains (β-gelatin) and three chains (γ-gelatins). The existence of multiple chain species indicates covalent bonds in the original tropocollagen molecules, hence the triple helices are not purely the products of physical interactions.

In the laboratory, however, much milder and more sophisticated processes are feasible, as in the preparation [153] of gelatin from rat tail collagen, and the samples produced are much less polydisperse. In some cases fractionation to achieve separate α, β and γ forms has been achieved.

Whether purified or not, gelatins dissolve in warm water, and above ~40 °C the molecules present are believed to be statistical coils [154]. On cooling such solutions below 40 °C, transparent elastic gels form spontaneously provided the concentration is high enough (>0.5%, for example), and this network-building process is now known to result from a frustrated attempt by the individual polypeptide polymers to return to the native collagen form. In the solution state this cannot be perfectly realised, and networks are generated which are cross-linked by limited triple-helical junction zones. The structural and rheological characteristics of these gelatin gels are now discussed. An excellent, if very early, review of the subject has been given by Ferry [3].

4.1.2 Structural Studies

Optical rotation (see Sect. 2.2.1) measurements have frequently been used in the study of gelatin gelation. An early example of such work is that reported by Smith in 1919 [155]. Since then many papers have appeared pursuing this approach, the results often being coupled with data from other measurement procedures such as viscosity and shear modulus determinations and NMR, dielectric relaxation and light scattering.

In 1948, in his review, Ferry described application of the optical rotation approach to gelatin solutions cooled below 40 °C, when two phases of optical rotation change were recognised. These were an initial phase, lasting several hours, and a much slower phase still progressing after days of storage. No detailed description of the molecular processes was given, however.

In 1960, Flory and Weaver [156] examined the optical rotation change occurring in gelatin solutions, rapidly quenched to temperatures in the range 5 to 23 °C. The gelatin concentrations were in the range 0.1 to 0.4 gm/100 ml, and results were compared with traces showing the concomitant variation of reduced viscosity. Changes in viscosity were observed long after the bulk of the optical rotation change was over, and it was noted that the amount of reversion to native collagen was very much dependent on temperature, far less recovery taking place at higher quenching temperatures than at low. The main change in optical rotation was over in twelve hours, and Flory and Weaver did not comment upon any slow residual change, though this must have been present.

From their observations that half-lives for the gelatin conformation change were independent of concentration, Flory and Weaver concluded that the reversion process was first order, a finding which was surprising indeed in the light of the accepted triple helical nature of ordered species being formed. They also noted the high negative temperature dependence of the reaction rate constant, i.e. the fact that the optical rotation change was greatly slowed down by raising the quench temperature towards the collagen melting temperature.

In an attempt to explain the first-order concentration dependence within the concentration range studied, Flory and Weaver proposed a model for helix formation, which had as its rate determining step, the formation of helical segments in the proline-rich parts of the gelatin chains. These segments were supposed to have a transient existence, being able to transform rapidly to triple helix through inter-molecular reaction. Flory and Weaver also discussed the temperature dependence of their data within the framework of this model and presented a formula for the reaction rate constant in terms of the amount of cooling below the collagen melting temperature imposed by quenching. A mechanism of nucleation and growth was envisaged.

At about the same time as Flory and Weaver's work, in a similar examination of gelatin conformation change, Harrington and von Hippel [157] confirmed the concentration independence of specific rotation data for low gelatin concentrations. In addition to the expected rapid initial optical rotation change, they described a very slow mutarotation occurring over days rather than hours. They pointed out that, whilst the reaction rate was first order with respect to gelatin concentration, it was second-order with respect to the fraction of chain residues still unfolded, i.e. plots of the log of the rate of change of the specific rotation (i.e. log $d[\alpha]/dt$) versus log $([\alpha]_\infty - [\alpha]_t)$ were linear, with slopes equal to two in all cases.

In an attempt to rationalise these observations, Harrington and von Hippel proposed a very rapid initial folding of the proline-rich segments to an ordered conformation this being followed by a slower and more complete ordering of individual chains, and finally the slow association of the ordered chains to produce triple helices. This model, which also found support from data relating to collagenase attack on gelatin, is significantly different from the account of events given by Flory and Weaver. In the Harrington and von Hippel mechanism, the first-order gelatin concentration dependence was explained by making the conformational change essentially unimolecular, and allowing helix formation to take place by a side-to-side association of preformed helices at a later stage. The second-order kinetics of the transformation (in relation to the fraction of residues in a disordered state), were tentatively explained in terms of a nucleation and growth model similar in essence to the Avrami theory of crystallisation [158], but it was necessary to postulate a non-random distribution of helical nuclei. In Harrington and von Hippel's theory it is clear that the single helix is ascribed a stability not anticipated in the Flory-Weaver model, and this stability was explained in terms of water molecules becoming involved and stabilising the helix through hydrogen bonding.

The models of Flory and Weaver, and Harrington and von Hippel remained subjects for debate and investigation for many years after their publication and gradually, in the light of experiments on purified gelatins (e.g. gelatins separated into single chains (α), double chain (β) and triple chain (γ) components) they have become amalgamated into a somewhat different view of events. It is now believed to be unlikely that the single helix has a separate existence, and the need for the Flory-Weaver intermediate to explain unimolecular kinetics has become less pressing as such behaviour in dilute solution can be interpreted in terms of an intramolecular reversed triple helix formation. As concentration increases, intermolecular helix associations become more probable, and the order of reaction with respect to gelatin concentration increases. The second-order aspect of the phenomenon, in terms of untransformed chain segments, becomes explained in terms of nucleation, growth and annealing processes occurring with rate constants showing markedly different temperature dependences. At low quenching temperatures nucleation is fast, growth is slower and annealing processes occur very slowly. At higher temperatures, the growth process dominates, and much longer, more perfect, helical structures are achieved. Detailed discussions of these schemes have been given by Harrington and Rao [159], Harrington and Karr [160] and Yuan and Veis [161], and in general the nucleus is seen as involving three chains and a degree of hydrogen bonding. The early stages of gelatin aggregation are expected to involve substantial changes in optical rotation but little change in viscosity, whilst later processes have less effect on optical rotation, but produce considerable changes in rheology. This has been demonstrated by Eagland et al. [153] who combined optical rotation measurements with viscosity and NMR and dielectric measurements of molecular mobility. They found that the nucleated species involved very little change in molecular mobility, but that as growth of ordered regions occurred, chains became immobilised. Eagland et al., however, proposed intramolecular processes even at higher concentrations, and proposed gel cross-linking through association of ordered species. Though some electron microscope evidence [162, 163] appears to support this theory, it is not currently accepted; instead a transition to inter-chain triple helix formation at higher gelatin concentrations is envisaged. This transformation from intra to inter-molecular chain ordering has been discussed by Finer et al. [164].

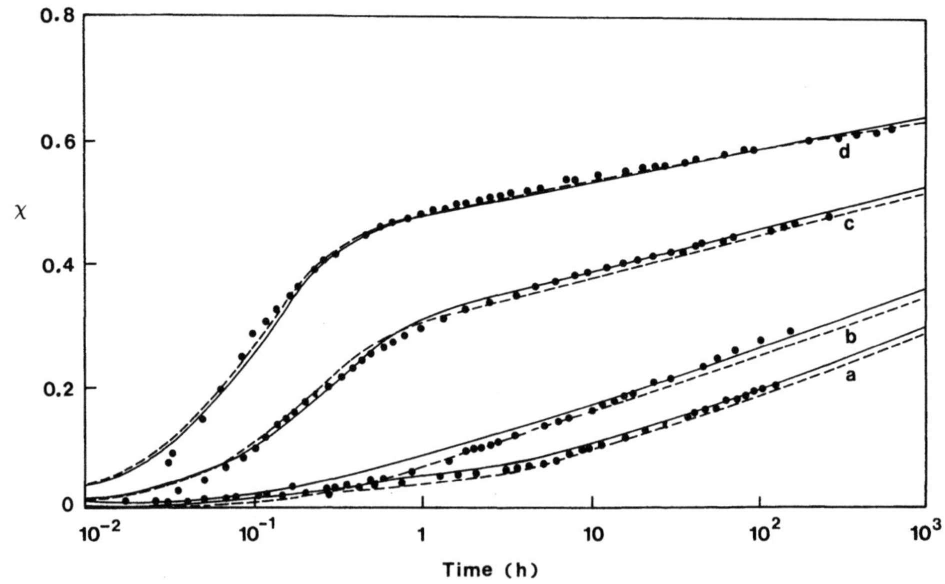

Fig. 24. Helix content (χ) of gelatin solutions (estimated from optical rotation data) plotted against logarithm of setting time (hours) at quench temperatures; (a) 28 C; (b) 26.5 C; (c) 20 C; (d) 10 C. Gelatin concentration 4.6% w/w. Continuous and broken lines represent predictions of two two-process growth models for helix development. Reproduced with Permission from British Polymer Journal, *17*, 169, 1985, Fig. 2, p. 171

Finally, in relation to optical rotation and gelatin gelation, recent work by Djabourov et al. [165,67], and Durand and co-workers [166,167] is of interest. Djabourov et al. have discussed the kinetics of optical rotation change for quenched gelatin solutions and have used a combination of Avrami crystallisation kinetics, and a longer time, logarithmic time dependence formula, to fit the data. They observe trends in the relative proportions of the two processes in transitions conducted at differing degrees of supercooling, the Avrami phase dominating at lower temperatures (Fig. 24). These processes have been identified by Djabourov and co-workers as corresponding to nucleated polymer crystallisation (Avrami exponent equal to unity) coupled with a secondary crystallisation at longer times.

This last view of gelatin organisation in solution agrees well with the results of a differential scanning calorimetric study by Godard et al. [168], and with recent work by Durand and co-workers [166,167]. These latter have measured the helix content of gelatin solutions as a function of time after quenching, using both optical rotation and ultrasonic absorption measurements. Four gelatin concentrations were examined (1.3, 2, 6 and 11% w/w) and several quenching temperatures between 36 °C and 10 °C. The gel point was identified in each case by viscosity measurements, and it was shown that, at any temperature, the fraction of gelatin in helical form at the gel point (h_g) was constant [166]. It was also shown that a plot of h_g against dilution C^{-1}, was linear, i.e. that h_gC was constant. It was therefore concluded that, at the gel point, in the gelatin systems studied, the critical helix concentration was independent of both gelatin concentration and quenching temperature.

Although Durand et al. have not discussed the implications of this finding in detail, h_g, the fraction of helix at the gel point, must indicate the critical cross-linking threshold appropriate to a particular sample concentration, whilst the linear C^{-1} dependence of this quantity must indicate increasing cyclisation and wasted triple helical cross-links as dilution is increased (as suggested in the Jacobson-Stockmayer model for ring-chain competition) [169].

Durand et al. have also examined the gel times in the light of their sensitivities to changes in concentration and temperature [167] and have, like Djabourov et al., identified two types of behaviours; short gel times and low sensitivity to temperature changes at low quenching temperatures, and high sensitivity and much longer gel times occurring as quenching temperatures are raised (>25 °C, for example). As in the work of Djabourov et al., the kinetics of optical rotation change at various temperatures, were fitted using the Avrami model. This fitted the low temperature data best, a rather different kinetic function becoming dominant at the higher temperatures. Again, a nucleated crystallisation process appears to combine with a secondary crystallisation (or annealing) process the exact blend in any situation depending on quench temperature.

4.1.3 Viscosity Measurements

Studies of optical rotation change in gelatin solutions, quenched to a given temperature, have often been accompanied by measurements of rheological properties such as reduced viscosity [3]. Generally it has been concluded that the timescales of such changes have been much longer than those appropriate to processes giving rise to initial changes in rotation. Yuan and Veis, for example, have described [161] a scheme for gelatin renaturation which contains processes which give rise to optical rotation change only (nucleation); processes which give rise to both optical rotation and viscosity changes (growth of helices); and processes which give rise mainly to viscosity changes (annealing). At higher concentrations it is to be expected that these processes will also have implications for the time course of development of the viscoelastic shear modulus. This is now considered.

4.1.4 Small-Deformation Rheology

The formation of triple helices is regarded as a principal source of cross-linking in gelatins, and hence of their elastic gel properties at suitably high concentrations.

The early stage of modulus growth during cross-linking has been examined both experimentally and theoretically by Peniche-Covas and co-workers [60] (see also Sect. 3.4). These authors made modulus measurements as a function of time through the gel point, and applied the cascade theory of network formation [112] in a quantitative treatment of the data. This approach which, up until Peniche-Covas et al.'s study, had been principally applied to the growth of covalently-linked synthetic polymer networks, was shown to be applicable to physically cross-linked biopolymer networks in a solvent (water). In the gelatin case the concept of functionality (number of potential cross-linking sites per aggregating unit) was generalised to include both the triple-

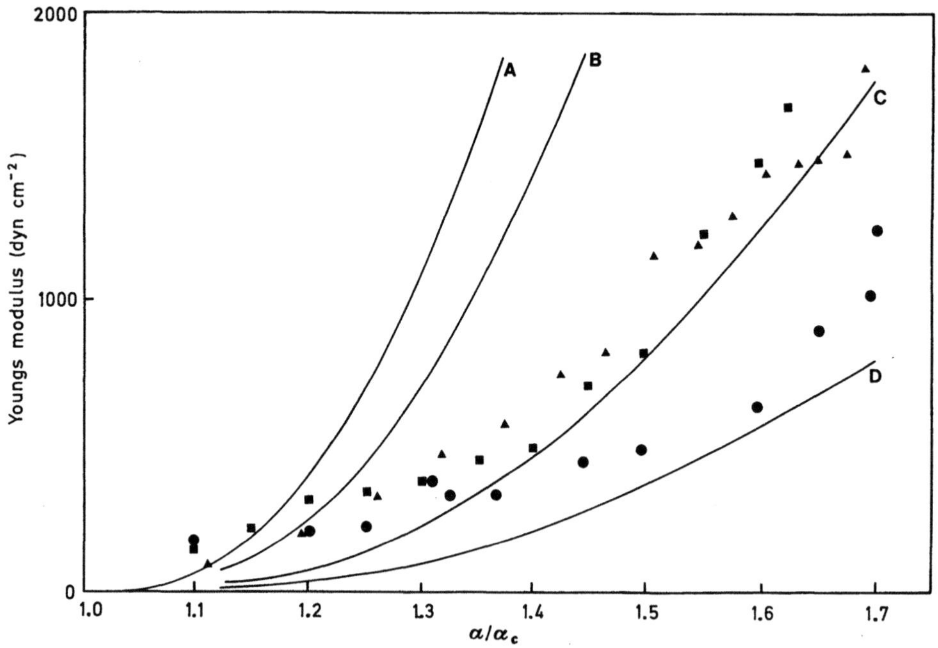

Fig.-25. Young's modulus E versus relative conversion α/α_c (from optical rotation) for gelatin solutions (three runs) quenched to 26.9 C. Solid lines depict theoretical predictions based on an ideal rubber theory (front factor = 1.0) and various specifications for the gelatin functionality. Curve C, which gives the best fit, assumes three gelatin chains per junction zone and eight zones per chain. Reproduced with Permission from Faraday Discuss. Chem. Soc., *57*, 165, 1974, Fig. 6, p. 177

chain nature of the junction zones and the likely number of such zones per polypeptide molecule. In the cascade calculation, information about the degree of cross-linking at the time t was supplied by optical rotation measurements. In this way the reaction coordinate α/α_c (the fraction of cross-linking sites used at time t, divided by the critical fraction at the gel point) could be evaluated as cross-linking progressed. By means of the cascade theory, and the Case-Scanlan definition [112] of an elastically-active network chain, the number of such chains per unit volume was evaluated as a function of time. Assuming an ideal rubber entropic contribution to the shear modulus of kT per chain, fits of modulus-versus-α/α_c data could be achieved. As is clear from Fig. 25 these were found to be sensitive to the functionality assumed for a gelatin chain, particularly to the value for the number of strands in the helix. The optimum value of three obtained in this work was consistent with the triple helical concept of the junction zones, and the fact that the ideal rubber description applied well to the network chains, was in accord with the usual picture of polypeptide polymers as being highly flexible. Gelatin networks, it appears, consist of triple helical junctions connected by sizeable lengths of disordered polypeptide chains.

Other publications have considered the time evolution of the modulus of gelatin gels. In such investigations, thermal history is important, and in a particularly detailed study [170, 171], te Nijenhuis has measured the growth of G′ and G″ for gelatin solutions, quenched rapidly to certain fixed temperatures. Variations in protein concentration, in quenching temperature, and in the frequency of modulus measurement were applied, and interesting results obtained. Temperature, for example, was found to have a very considerable effect on the cure curves, curves at lower temperatures showing a sigmoidal initial phase, with a linear long time region following (Fig. 26).

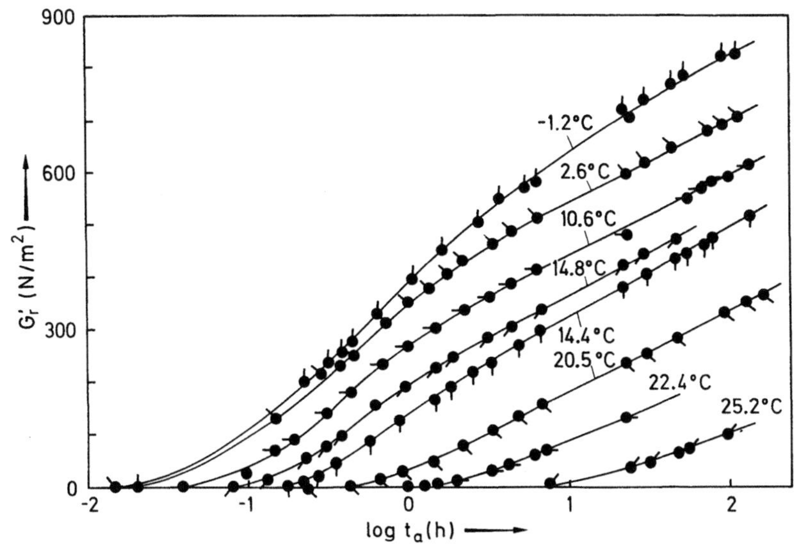

Fig. 26. G′ (measured at 0.393 rad/s) of gelatin solutions (1.95% w/w) plotted against logarithm of setting time (hours) for a series of quench temperatures. Reproduced with Permission from K. te Nijenhuis, Colloid and Polymer Science, *259*, 522 (1981), Fig. 7, p. 527

At high temperatures the phase of linear modulus increase became dominant. Application of a kinetic model to the cure curve data suggested a third-order process, and this was considered to be consistent with intermolecular triple helix formation. The temperature induced shift in form of the cure curves conforms well with the kinetic description of optical rotation behaviour discussed earlier, and with the DSC evidence of Godard et al. [168]. The link between modulus and optical rotation, however, is still under investigation, e.g. by Djabourov and co-workers. It is worth adding that te Nijenhuis found little frequency dependence for G' and G'' for the systems studied.

Despite difficulties in identifying equilibrium states for gelatin gels, attempts have been made to establish a relationship between modulus and concentration. As early as 1948, for example, Ferry [3] discussed this matter and provided evidence for a C^2 dependence of G'. He rationalised this in terms of a cross-linking equilibrium involving pairwise reaction of cross-linking sites, and this approach was used later (see Sect. 3.5) to provide a relationship between gel melting temperature and polymer concentration and molecular weight (Eldridge-Ferry theory) [84]. It is interesting that in these early studies the triple chain model for gelatin junction zones was unknown.

Ferry's work was largely concerned with gelatin gel moduli at concentrations well above critical. He noted, however, that the C^2 law did not hold at the lower concentration end of the scale. This point has recently been examined by Clark et al. [137] and Clark and Ross-Murphy [70] who established an operational definition of the gelatin modulus in terms of a curing time and measurement frequency. At high concentrations the C^2 dependence for G' was confirmed, but a much higher power law was obtained at concentrations much nearer the critical value. In a quantitative fit to the G' versus concentration data, Clark and Ross-Murphy exploited a universal relationship [35] derived originally for globular protein gels (Sect. 5.3.1). This function, and details of its application to experimental data (such as the gelatin data) have already been described (Sect. 3.4).

4.1.5 Ultimate Properties

The rheological aspects of gelatin gel formation, just discussed, have related mainly to measurements involving small strains. Data also exist in the literature describing the behaviour of such gels at larger strains, including their failure. In recent articles by McEvoy et al. [95, 96], for example, large deformation test experiments have been described in which gel rings, moulded from quenched gelatin solutions, were subjected to extensional strain, and stress-strain curves recorded up to failure. Various gelatin concentrations were used in this work and various strain rates applied. The stress-strain curves obtained were used to derive values for Young's moduli from initial slopes, and these were found to be consistent with previous modulus data from mechanical spectroscopy if the relation $E = 3G$ was assumed. In addition, the form of the stress-strain curve was found to be insensitive to strain rate at a given concentration, and could be mathematically modelled using the BST equation (Sect. 3.3.1). Stress-to-break and strain-to-break values were highly strain rate dependent, however, and such data were handled using a failure envelope approach. Envelopes derived for the gels at different concentrations could be scaled to form a composite master curve, and the scale factors necessary for this exercise were found to show the same concentration dependence as the linear viscoelastic shear modulus.

4.2 Polysaccharide Gels

4.2.1 Marine Polysaccharides

4.2.1.1 Carrageenans

In any discussion of the gelling properties of marine polysaccharides it might seem natural to commence with agar for this material must surely be the most familiar example of such gelling substances. However, examination of current literature suggests that it would be more profitable to begin with a description of the properties of the carrageenans, a group of polysaccharides also derived from red seaweeds, and closely related to agar structurally.

The carrageenans [172] are sulphated polysaccharide polymers which can be extracted from various genera of marine algae of the class Rhodophyceae. They are usually unbranched, and are based on an AB disaccharide repeat unit (Fig. 27), i.e. they can

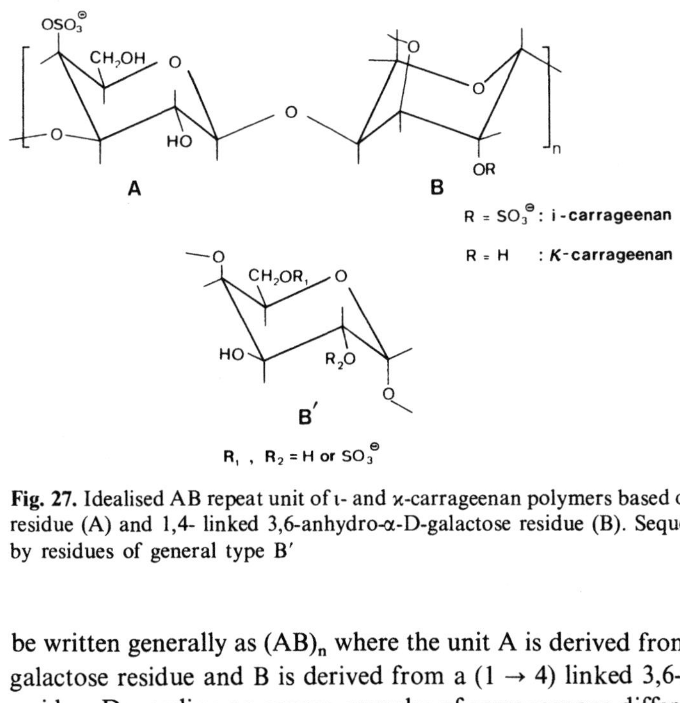

Fig. 27. Idealised AB repeat unit of ι- and ϰ-carrageenan polymers based on 1,3- linked β-D-galactose residue (A) and 1,4- linked 3,6-anhydro-α-D-galactose residue (B). Sequence is broken occasionally by residues of general type B′

be written generally as $(AB)_n$ where the unit A is derived from a (1 → 3) linked β-D-galactose residue and B is derived from a (1 → 4) linked 3,6-anhydro-α-D-galactose residue. Depending on source, samples of carrageenans differ in the extents to which they carry sulphate groups, and in the positioning of these substituents along the polymer chain. Thus, in practice, the formulation $(AB)_n$ is an idealisation of the real situation. Nonetheless it is possible, by selecting suitable genera of algae, to extract at least two classes of carrageenan which conform closely to the simple repeat formula. These materials are known as ι and ϰ-carrageenan, and as Fig. 27 makes clear, in ι carrageenan, rings A and B both carry sulphate groups, whilst in ϰ-carrageenan only ring A is substituted. Although other types of carrageenan exist, including the more disordered forms already referred to, most research has been concentrated on the gelling properties of ι and κ-carrageenan, and in what follows, attention will be focussed on these two polysaccharides, exclusively.

Whilst carefully selected and purified samples of ι and κ-carrageenan do conform closely to the ideal formulations represented in Fig. 27, they are always found to deviate from ideality in one important aspect. This pertains to the anhydro-galactose ring B which is found to be partly replaced (albeit infrequently) by (1 → 4) linked galactose-6-sulphate and 2,6-disulphate (Fig. 27). These residues are considered to be biological precursors of the anhydro-form, and as will emerge in the discussion below, they are believed to play an important part in the assembly of carrageenan polymers into network structures.

In seaweeds, hydrated carrageenan networks are believed to be of structural importance, but the importance of carrageenans to man is in their ability to form thermoreversible gels. Hot solutions of carrageenans, often containing quite low concentrations of polymer (~1% w/w), tend to form rigid gels on cooling, and this property finds practical applications particularly in the food industry. Scientific studies of carrageenan gels, intended to broaden the range of such applications and improve their efficiency, have largely been concerned with identifying the molecular mechanisms of network formation when hot mobile solutions are cooled, and the first clues as to what might be involved in the cross-linking of carrageenan polymer chains came from fibre diffraction [173)]. In this early work gelling solutions were cooled and stretched, and examined by standard fibre diffraction methods. Various ion forms of ι and κ-carrageenan were considered, and diffraction patterns were obtained in most cases, particularly good results being obtained using potassium ions. It was concluded that the ι-carrageenan system contained three-fold right-handed double-helices stabilised by hydrogen bonding, but the situation was less clear for κ-carrageenan. Similarities between the patterns obtained for ι and κ-carrageenan, however, led to the proposal that for κ-carrageenan a related double helical structure was involved.

In the original fibre diffraction paper [173)] it was suggested that the ι-carrageenan double helix might well constitute the basis of cross-linking in ι-carrageenan gels. It was also suggested that multiple associations of the double helices might also be involved, and the fact that a gel network formed, rather than a precipitate, was ascribed (at least in part) to the presence of sequence breaking residues ('kinking' residues) of the galactose sulphate type referred to earlier.

In the years following, much research was carried out to establish the truth of this proposal. As in the case of gelatin aggregation, a large part of the effort was concentrated on optical rotation studies [174)]. On cooling carrageenan solutions a large optical rotation change takes place (Fig. 28) and application of the semi-empirical rule developed by Rees [175)], and referred to earlier (Sect. 2.2.1), made it possible to predict the amount of optical rotation change to be expected if conversion of a random coil to a three-fold double helix were really involved. Early measurements checked well with this theory and, provided that the conversion from coil to helix was very substantial during gelation, the evidence obtained was highly supportive of the double helical model. It should be noted that the change in optical rotation, measured at any one temperature for ι-carrageenan, was extremely rapid, so that unlike in the gelatin case optical rotation-versus-temperature profiles were studied, rather than optical rotation-time curves at given temperatures.

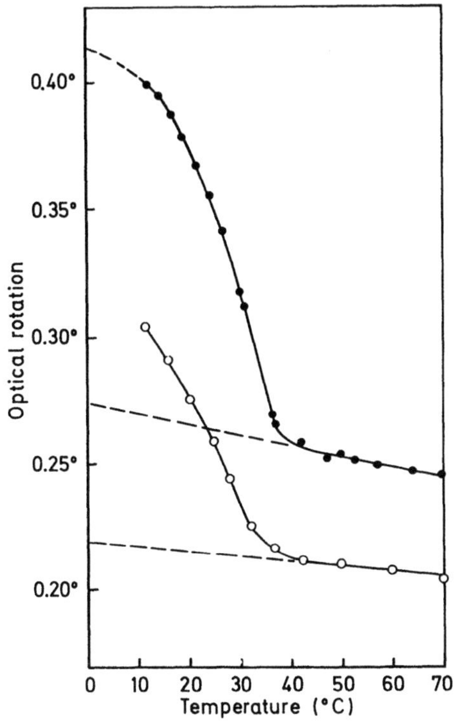

Fig. 28. Changes in OR (436 nm) as solutions (6% w/v) of native ι-carrageenan (○) and ι-carrageenan segments (●) are cooled. Results independent of thermal history. Reproduced with Permission from Faraday Discuss. Chem. Soc. *57*, 221 (1974), Fig. 4, p. 226

The gel systems themselves were less than ideal for studying optical rotation effects (possibilities of birefringence, light scattering etc.) and so, to check conclusions, experiments were repeated using so-called carrageenan segments [176, 136]. These were obtained by cleaving the native polymers at their 'kink' points (e.g. points of galactose-6-sulphate substitution). Optical rotation studies of these more idealised materials (which did not gel under any conditions) confirmed that a very substantial increase in rotation occurred as solutions were cooled, and that the amount of this change was in agreement with Rees's prediction for total conversion to double helix. Both ι and ϰ-segments were studied, and for ι-carrageenan, the position was particularly simple, as the optical rotation responded rapidly as temperature was altered, and the changes observed were reversible [176]. For ϰ-carrageenan segments, the situation was more complicated [136] as a pronounced hysteresis occurred (Fig. 29), there being apparently some inhibition of formation of the ordered state on cooling.

For ι-carrageenan, equilibrium was assumed to be achieved by the segments, at all temperatures, and questions were asked about the sizes of the molecular weight changes accompanying the optical rotation transition. Early investigations by Jones et al. [177] using osmometry and light scattering suggested strongly that both the weight and number average segment molecular weights doubled during the cooling process. For ϰ-carrageenan segments, no very definite answer was obtained, but much larger changes in molecular weight were involved. For ι-carrageenan, the molecular weight doubling was exactly what would have been expected for the formation of matched double helices, and appeared to confirm the double helical model for gel junction zones.

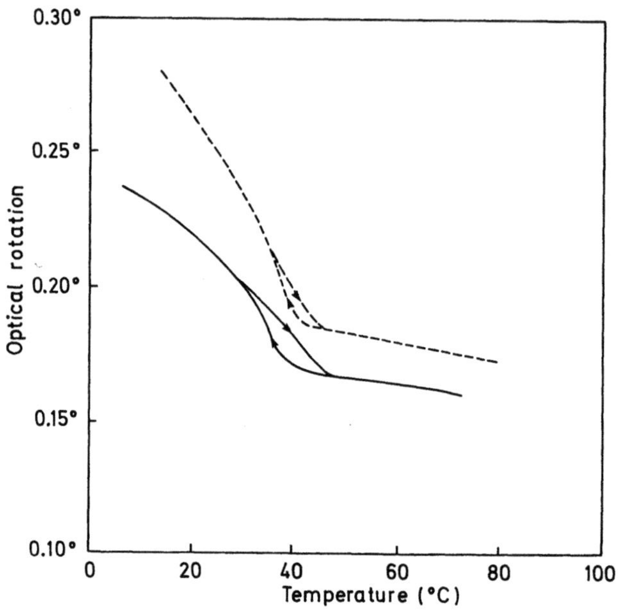

Fig. 29. Changes in OR (546 nm) as solutions (3% w/v) of native (————) and segmented (— — — —) ϰ-carrageenan are cooled and reheated (see arrows). Reproduced with Permission from J. Molecular Biology, *68*, 153, (1972), Fig. 3, p. 160

In subsequent studies of segments, evidence from other techniques such as thermodynamics [178], NMR [25] and fast reaction kinetics [179–182] has also been supportive of double helix formation by ι-carrageenan in solutions and gels. Comparison of calorimetric heats of reaction (from DSC) and van't Hoff heats from optical rotation-based equilibrium constants at first suggested [178] that the process of coil-to-helix transformation was far from cooperative, but inclusion of known facts about segment polydispersity allowed a highly cooperative description to be rescued, and all experimental data explained. A two state all-or-none mechanism was also indicated by ^{13}C NMR studies of ι-carrageenan segments in solution [25], for at temperatures where the optical rotation indicated complete conversion to helix, all high resolution NMR signal disappeared (Sect. 2.2.3), and at intermediate stages, there was no indication of a transient exchange between helix and coil segments. Finally, and more recently, kinetic studies of the optical rotation change by fast reaction techniques [179–182] have also supported cooperative double helix formation. The change was induced by a salt-jump procedure and measured in a stopped-flow polarimeter. This approach by Norton and co-workers gave a clear indication of a second-order rate-limiting process involving the nucleation of a two-chain aggregate structure. Evidence for initial double helix formation in solutions of ϰ-carrageenan segments was also obtained by this technique.

By 1980, as a result of the carrageenan segment investigations, the double helical model for junction zones in carrageenan networks was firmly established. Some problems remained, however, principally connected with well established observations about cation effects on gelation. It was known, for example, that both ι- and ϰ-carrageenans formed strong gels in the presence of K^+, Rb^+ and Ca^{++} ions, but that there was little gelation in the presence of Li^+, Na^+ or $(CH_3)_4N^+$. The segment studies conducted earlier had not addressed this issue, being conducted under restricted conditions of electrolyte content.

The counterion issue in carrageenan systems is an important one, and in the last few years many experiments have been performed to explore it, and clarify the situation. Work done in 1980 by Morris and co-workers [183] is particularly revealing for, by a series of experiments, some essential aspects of the problem were investigated. Optical rotation studies of the aggregation behaviour of tetramethyl ammonium ι-carrageenan segments (and the native carrageenan in this salt form) showed clearly that helix formation was taking place, but in the unsegmented polymer solutions no gelation occurred. This eliminated any argument that poor gelling in carrageenans could be linked to a lack of formation of the ordered form. Similar studies of Li^+ and Na^+ and K^+ salt forms showed also that gel strength could in no way be connected with the extent of optical rotation change. The Li^+ and Na^+ native i-carrageenan solutions, for example, produced rather weak gels, whilst under the same conditions of concentration and degree of ordering, the K^+ gels were much stronger.

Clues to the solution to this problem came from light scattering on both native material, and segments [183]. In the presence of $(CH_3)N^+$, and Na^+ ions, segments showed only a doubling of molecular weight, whilst when K^+ was present, the molecular weight was much higher. In addition, for the native $(CH_3)_4N^+$ material, which did not gel, a limited amount of aggregation was detected by light scattering, and this, and the other observations, led Morris and co-workers [183] to modify the existing description for carrageenan networks by proposing a 'domain' model. Here, the idea was that junction formation occurred partly by double helix formation and partly by helix-helix association, rather than totally by double helix formation, as had been assumed previously. In the new model it was proposed that the double helical mode of association occurred only to a limited extent, and that association of the resulting 'domains' then relied upon specific cation effects for their subsequent promotion (Fig. 30).

Since 1980 the study of specific ion effects on the conformational properties of ι- and κ-carrageenans has continued, and quite recently has been extended to include anion effects. Optical rotation has continued to be used as a principal investigative technique, but cation and anion NMR methods have been introduced to look for specific ion binding. Evidence for the binding of Na^+, K^+ and Rb^+ cations to the sulphate groups of the carrageenan ordered form has been obtained by the NMR method [184], but Norton et al. [185] also report binding to coil forms prior to the disorder-to-order transition. In network formation by polysaccharides specific binding effects may be involved, but other explanations of ion specificity such as influence on water structure [184] cannot be ruled out. This latter explanation has been advanced by Norton et al. [186] to explain anion effects on carrageenan conformation, such as a tendency for certain anions to promote helix formation, without at the same time promoting helix-helix association. Smidsrød et al. [187], however, believe that a specific interaction may be involved.

Anion effects on carrageenans are of interest for another reason, for studies of samples of ι- and κ-carrageenans, in the presence of iodide ions, have led Smidsrød and co-workers [188, 189] to propose that the ordered conformation produced by cooling,

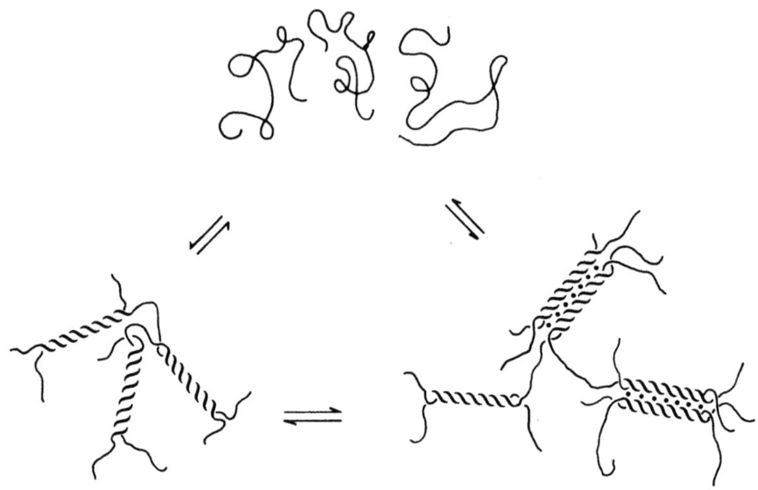

Fig. 30. Domain model (ref. [183]) for ι-carrageenan network formation. Cross-links involve both double helix formation and limited association of double helices (with possible counterion involvement)

and observable by optical rotation change, is a single, not a double, helix. This conclusion was based on molecular weight measurements by osmometry, for no molecular weight change was found on passing through the disorder-to-order transition, and the formation of a single helix was further suggested by a measured concentration independence of the specific rotation-temperature curves. These findings have led Smidsrød et al. [189] to propose a network model for carrageenan gels (particularly

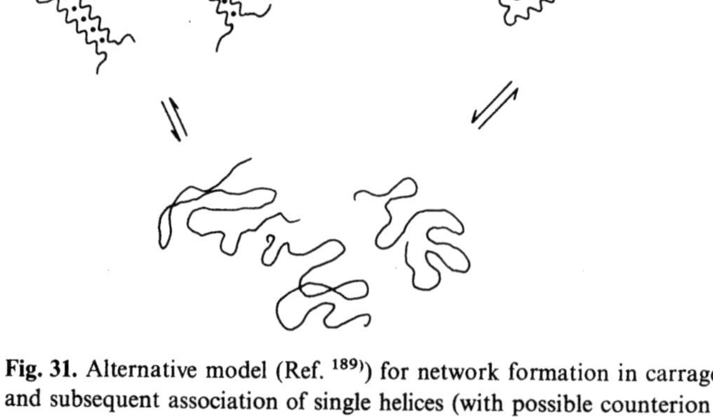

Fig. 31. Alternative model (Ref. [189]) for network formation in carrageenans based on development and subsequent association of single helices (with possible counterion involvement)

for ϰ-carrageenans) involving the association of single chains (in ordered conformations) to form dimeric junction zones, with cations probably also involved (Fig. 31). Norton et al. have disputed this model [186], as their studies of the carrageenan-iodide system by kinetics and light scattering still support the double helical description.

The discussion so far, of carrageenan gelation, has been largely structural, and has been concerned principally with the mechanism of cross-linking, and hence with structural features extending over very short distance scales. Rheological properties of the networks have rarely been mentioned, and when they have had a bearing, purely qualitative remarks have been made, such as to describe gels as weak or strong. This bias toward microstructure has arisen, however, because, up until recently, the literature of carrageenan research has contained very little rigorous, quantitative, rheological data. Although much remains to be done, the situation with regard to carrageenan rheology has improved slightly of late.

V. J. Morris and co-workers [184, 190–193], for example, have made shear modulus measurements on both ι- and ϰ-carrageenan gels and have plotted modulus-concentration relationships for gels made in the presence of different cations (Fig. 32). They have also considered the effect of temperature on modulus results. For ι-carrageenan gels this work has quantitatively confirmed the positive promotion of gel rigidity by K^+, Rb^+ and Cs^+ ions, but it has also shown that gels can arise in the presence of Li^+ and Na^+ under forcing conditions (high polymer and electrolyte

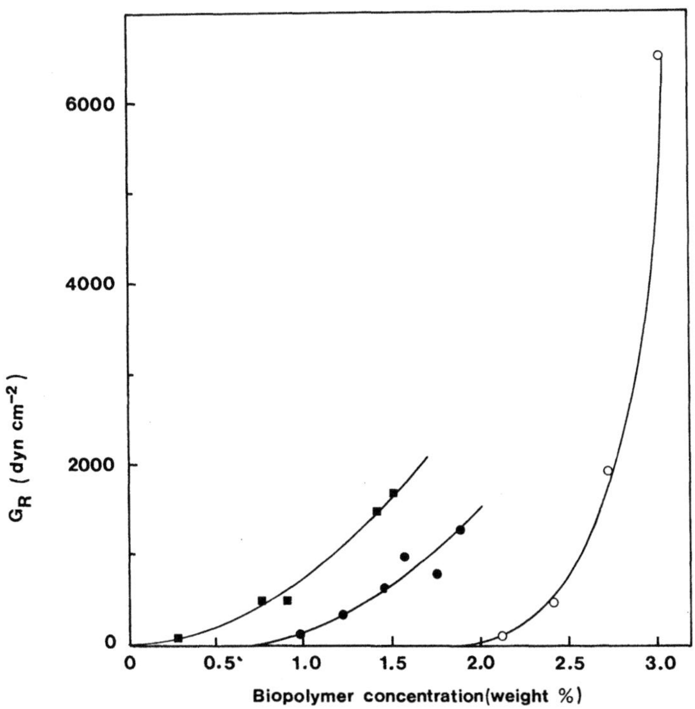

Fig. 32. Rigidity modulus G (measured in U-tube apparatus) versus-concentration data for gels based on calcium (■), potassium (●) and sodium (○) ι-carrageenate. Reproduced with Permission from J. C. S. Chemical Communications, 983, 1980, p. 983, 1980

concentration). In addition, it has demonstrated that Ca^{++} ions influence both ι- and ϰ-carrageenan to form strong gels, and this finding seems in some measure to contradict the traditional notion that ι-carrageenan is 'calcium sensitive', and ϰ-carrageenan 'potassium sensitive'.

As a result of their modulus-concentration work, Morris et al. have emphasised [184] the importance of the entanglement concentration C* in determining polysaccharide network formation, for they contend that gels form only from entangled solutions. This identification of the critical concentration with C* is strictly at variance with the observations by Clark and Ross-Murphy [70] (Sect. 3.4) on the behaviour of globular protein gels, for critical concentrations in the globular protein gel case were always found to be much less than C*. In general, as suggested earlier, for finite energy cross-links C_0 must be $<C^*$, because in this case substantial overlapping of chains is not required to produce gelation.

V. J. Morris et al. [191–193] have analysed their modulus concentration relationships using an ideal rubber theory; and have derived values for M_c the molecular weight between cross-links. They argue for long Gaussian chains close to the critical concentration, and for shorter network chains at higher concentrations, but the ideal rubber model they adopt may be contested in the context of polysaccharide gels particularly at concentrations significantly above critical, because of the established conformational inflexibility of the polysaccharide backbone [101].

4.2.1.2 Agar

Like the collective term, carrageenan, the name agar refers to a complex mixture of polysaccharide components which may be derived from certain genera of the Rhodophyceae group of red sea weeds. The principal gelling component in the mixture is agarose, and like the carrageenans, this substance is based on a disaccharide repeat unit (Fig. 29). Unlike the carrageenan case, however, in agarose (and in its derivatives) the 3-linked-β-D galactose residue (A) is connected to a 4-linked 3,6-anhydro-α-L-galactose residue (B) i.e. to the L, not to the D, galactose anhydride. Molecular weights of agarose polymers have been estimated to be $\sim 10^5$.

Although agarose is the principal component of commercial agar, other species present in both raw agar and in commercial preparations, include material derived by substituting the β-D-galactose residue at the 6 position (the 6 methyl ether and 6-sulphate, for example) and material obtained by substituting the anhydride at the 2-position (the 2-sulphate derivative).

In addition to these impurities, or extra components, in agar, chains of agarose residues can deviate from the ideal repeating $(AB)_n$ structure of Fig. 29, by having some of the 3,6-anhydro-α-L-galactose residues replaced by galactose itself or by galactose-6-sulphate. As in the carrageenan situation such 'kinking' residues enable the agarose chains to be broken into segments by Smith (periodate) degradation [136], and again, as in the carrageenan case, they are believed to play an important role in the gelation mechanism of agarose, by preventing perfect ordering, and hence precipitation, of agarose chains during the sol-gel transformation. The agarose segments, however, have not played such a large part in studies of agarose gelation, as they have done in the carrageenan case. The absence of charged sulphate groups causes them to be precipitated from solution on cooling.

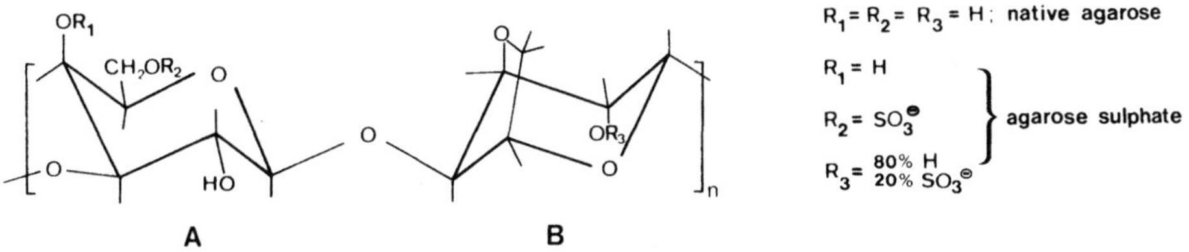

Fig. 33. Idealised AB repeat unit of agarose polymer based on 1,3- linked β-D-galactose residue (A) and 1,4- linked 3,6 anhydro α-L-galactose residue (B). Possible patterns of substitution involving sulphate groups are also indicated

The thermoreversible gelation of agarose is a well-studied process and occurs when hot agarose solutions are cooled below ~40 °C. In the hot solution state agarose molecules appear to behave as stiffened coils [194], but when such solutions are cooled, stiff, turbid, brittle gels are formed at concentrations of polysaccharide in excess of 0.1% w/w.

Agarose gels show both syneresis and hysteresis, i.e. they tend to exude water on standing, and the gel 'melting point' is much higher (~90 °C) than its 'setting point' (~40 °C). In fact, agarose represents an extreme of behaviour in the series ι-carrageenan, ϰ-carrageenan, agarose; and this may be rationalised in terms of its (ideally) almost zero content of ionisable substituents, and the increasing sulphate content shown by the other two.

As for carrageenan gels, much attention has been paid to the molecular origins of cross-linking in agarose gel networks. Work by Arnott et al. [195], in 1974, using X-ray fibre diffraction suggested that double helix formation might be involved, and the fibre diffraction evidence coupled with optical rotation data suggested that for agarose a left-handed (not right-handed as for ι-carrageenan) threefold helix was present. Again it turned out that changes in optical rotation, occurring during the sol-gel transformation, could be explained in terms of the ordered conformation in the fibre, provided that the helix content of the gels was high.

From evidence from other techniques, however, it seems, that the agarose network structure involves more than just double-helix formation. Evidence for a substantially aggregated structure for the junction zones comes both from the hysteresis property [196] evident in optical rotation-temperature plots [136] (Fig. 34), and from light scattering experiments [54–56, 197, 198], electron microscopy [199] and gel permeation chromatography [200]. Substantial bundles of agarose chains (Fig. 35) forming gel junction zones have also been indicated by the results of viscometric and fluorescence depolarisation studies [201, 202]. It is to be noted that in this work the agarose gel formation was described in terms of Flory's phase diagram for increasing concentration of rods (Sect. 3.7). This implies that anisotropic microcrystalline junction zones (based on aggregated helices) are dispersed through an isotropic supporting phase, and suggests that in agarose gels only a limited proportion of the agarose is in an ordered state. Since such a conclusion contradicts at least one of the basic assumptions which led optical rotation data to be interpreted in double helical terms, this work must presumably be regarded as providing evidence against the double-helix hypothesis.

In fact, where agar gels are concerned, the double helix junction zone concept has indeed been challenged. Foord, for example [203], has suggested the presence of single

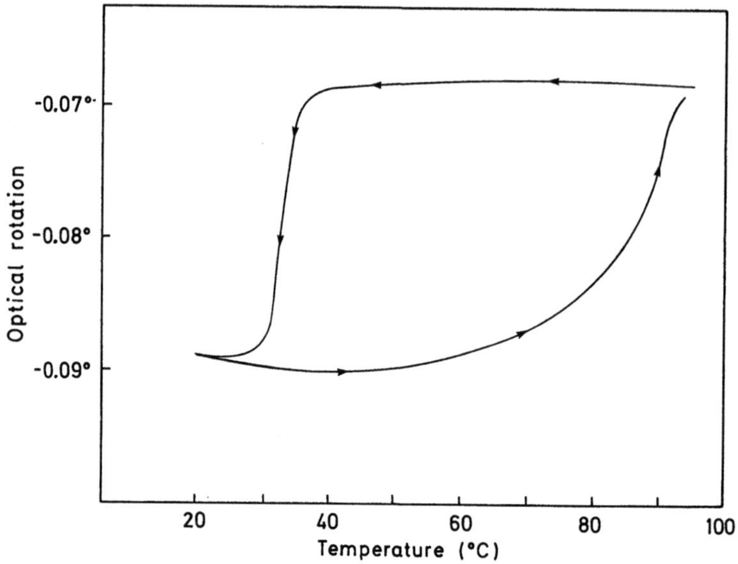

Fig. 34. Changes in OR (546 nm) as solutions (0.2%) of native agarose are cooled and reheated (see arrows). Note substantial hysteresis. Reproduced with Permission from J. Molecular Biology, *68*, 153 (1972), Fig. 4, p. 160

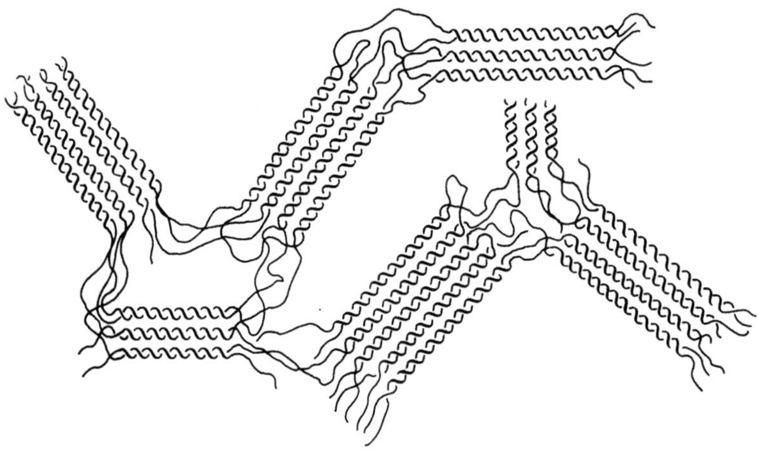

Fig. 35. Model for agarose network formation. Cross-links involve both double helix formation and substantial association of double helices to form microcrystalline junction zones

helices as a result of X-ray diffraction and has proposed a model for the agar network structure based on fibrillar crystals containing these units. This is similar to an earlier description of gels based on synthetic polymers by Atkins [204]. It seems that for agarose gels, as in the carrageenan case, there is still some controversy about the exact molecular conformations involved in the cross-linking phenomenon.

The process of network formation in agar gels, whether it involves the formation of bundles of double, or single, helices is a disorder-to-order phenomenon involving polymer association. Light scattering studies of this process by Prins and co-workers [54, 55, 198], which included kinetic experiments, have led to the suggestion that gel networks formed by sudden quenching of agarose sols arise via the non-nucleated mechanism of spinodal decomposition [150]. A nucleated mechanism of network formation may occur in other thermal circumstances, however. Where the spinodal mechanism is relevant, it allows agarose networks to be described as microphase separated structures involving a periodic variation of network density over a short distance scale.

Whilst the structural properties of agarose gels have been investigated extensively, the same is not quite so true of the rheological implications of agarose network formation. Some work has been carried out, however, both at small and large deformation. Clark et al. [137, 70], for example, have made measurements of the shear modulus of agarose gels at various concentrations, and have shown (a) that the critical concentrations C_0 is very low ($\sim 0.2\%$ w/w), (b) that the power law dependence of the modulus on concentration tends to the limiting form $G' \propto C^2$, at the higher concentrations accessible, but is $G' \propto C^n$ at lower concentrations (closer to C_0) with n very much greater than two, and (c) that the contribution of individual elastically active agarose network chains to G' is much greater than would be predicted purely on entropic grounds, and is much higher than that produced by the active chains in gelatin gels. The shorter and stiffer segments of single agarose chains in agarose gels at higher concentrations appear to give rise to this effect.

Agarose gels have also been studied rheologically at higher deformation by McEvoy et al. [95, 96], and their very brittle character related to structural features of the type just mentioned. The methodology employed in this work was similar to that used in related studies of gelatin gels described earlier (Sect. 4.1.5).

4.2.1.3 Alginates

Alginates are salts of alginic acid and occur as intercellular material in brown algae [205]. Their function as structure-forming components of these marine plants appears to result from their capacity to gel in the presence of certain divalent cations.

In molecular terms, alginic acid is composed of two distinct types of monosaccharide; i.e. 1,4-linked β-D-mannuronic acid (M) and 1,4-linked α-L-guluronic acid (G), these residues being present in varying proportions depending on the alginic acid source. It turns out that alginic acid (and salts) is a block co-polymer containing both MM... and GG... homopolymer blocks and mixed blocks containing irregular sequences of M and G units. The M and G residues and their modes of linkage are shown in Fig. 36.

A structural description of alginates in purely sequential terms is clearly difficult to specify statistically as, apart from specification of the M/G ratio of a particular material, there will be details of block length and block arrangement to be considered. Chemical treatments [206, 207] do allow alginates to be degraded to blocks containing MM... and GG... sequences, however, so these can be characterised as separate entities. For these segments the average block length is usually greater than twenty residues.

Unlike the carrageenans and agar, alginates do not form thermoreversible gels. Instead, network formation is induced in solutions of the sodium salt, when divalent cations such as Ca^{++} are introduced. Such replacement of Na^+ by Ca^{++} leads to

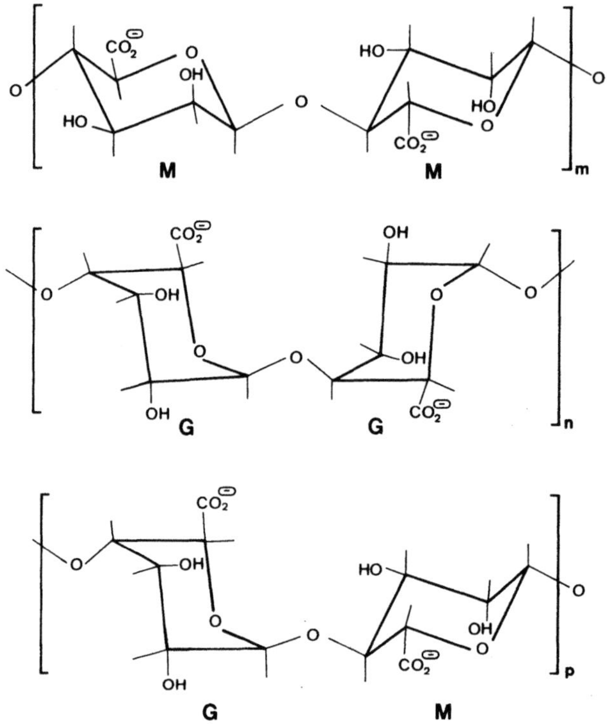

Fig. 36. Structures of mannuronate, guluronate and mixed blocks constituting alginate polymers. Mannuronate blocks are based on 1,4- linked β-D-mannuronic acid (M); guluronate blocks on 1,4-linked α-L-guluronic acid (G). Mixed blocks are linked as shown.

stiff, brittle (often turbid) gels, the exact rheological character, and turbidity, depending on the alginate type (e.g. the M/G ratio) and the mode of introduction of the divalent ions. Thus, Ca^{++} ions can be introduced by diffusion, using dialysis methods, or introduced *in situ* by adding inorganic calcium salts of low solubility, and controlling pH.

For the free ungelled alginate polymer, investigations using viscometry and light scattering [208–210] have shown that even at high ionic strength, the coils present in solutions are unusually expanded. The alginate chains are therefore intrinsically inflexible, and detailed studies of different block types, has suggested the order of inflexibility as

$$GG > MM > MG$$

and this is in accord with the results of conformational energy calculations [211]. A typical value [208] for the weight average molecular weight of alginate is $\sim 5 \times 10^5$.

Gelling of alginates in solution by the introduction of divalent ions has its origins in a capacity for specific ion binding accompanied by conformational change. Calcium ions, it appears, bind preferentially to guluronate (G) blocks, and this process has been shown [212,213] to be highly co-operative if the guluronate block exceeds twenty residues. No evidence for a similar high specificity or co-operativity has been found in the mannuronate case, mannuronate apparently showing standard polyelectrolyte behaviour.

Other physical studies have revealed differences between the guluronate and mannuronate sequences. Fibre diffraction studies [214,215], for example, of the acid form of alginates, alginic acid, have shown that whilst mannuronic acid takes up a two-fold ribbonlike helix in the acid form, and an extended three-fold helix in the salt form, guluronic acid sequences show somewhat different conformational behaviour. Only two-fold ribbon helices are formed by guluronic acid, or the guluronates, the diaxial 1–4 linkages of these ensuring a highly buckled structure, with a shorter repeat distance, and the distinctive feature of having well defined electronegative cavities capable (in the salt form) of accommodating cations of appropriate size.

Circular dichroism, as has been mentioned earlier (Sect. 2.2.1), has also played a large part in the physical characterisation of alginate sols and gels. Morris et al. [216] and Grant et al. [217] have observed very substantial changes in the rotatory strengths of circular dichroism bands corresponding to the $n \rightarrow \pi^*$ transition of the guluronate carboxylate residues, as Ca^{++} ions are added and gelation takes place (Fig. 37). A combination of this observation, fibre diffraction data, and molecular model building, led Grant et al. to propose an 'egg-box' model [217] for the junction zones in alginate gels. In this model guluronate sequences in alginates associate into matched crystalline aggregates in the two-fold ribbon-like form, ions co-operatively bound during the process, sitting inside the electronegative cavities, like eggs in an egg-box (Fig. 38). Subsequently Morris et al. [218] have, to some extent, modified this description, by limiting it (in non-forcing conditions of calcium concentration) to a side-by-side

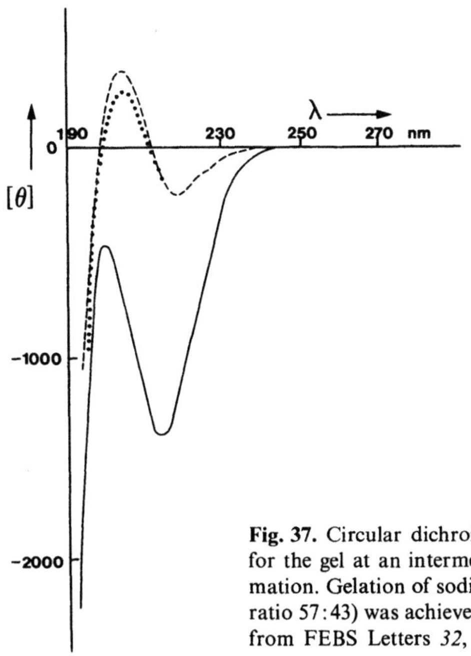

Fig. 37. Circular dichroism spectra for alginate solution (———) and for the gel at an intermediate (– – – –) and final (. . . . . .) stage of formation. Gelation of sodium alginate (0.1% w/w, mannuronate:guluronate ratio 57:43) was achieved by adding Ca+ +. Reproduced with Permission from FEBS Letters *32*, 195 (1973), Fig. 1, p. 196.

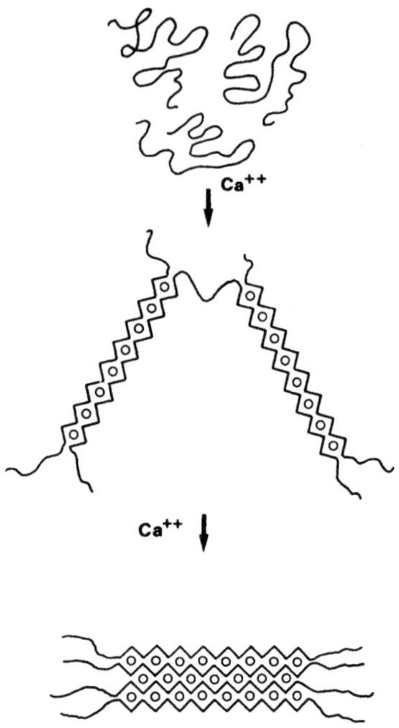

Fig. 38. Model for alginate network formation. Cross-links involve lateral association of guluronate segments to form ordered junction zones. Group II cations (particularly Ca++) occupy sites in these structures, and in the original "egg-box" model (Ref. [217]) combination of several chains was proposed. More recently, a model involving only the dimerisation of chains has been suggested (Ref. [218])

association of two chains only. This dimeric unit (see also Fig. 38), is proposed as a junction zone in alginate gels, analogous to the ion-induced association of carrageenan helices in carrageenan gels, and evidence for it comes from further experiments employing CD measurements, and ion-binding studies, involving equilibrium dialysis. Morris et al. also cite electron micrograph evidence [205] (originally presented by Smidsrød) in support of the dimeric junction zone model. The higher aggregates, it appears, are likely to occur only under forcing conditions of high Ca^{++} concentration. In extreme conditions there may also be some association of mannuronate sequences, but more usually these are the connecting links between junction zones.

The structural picture of alginate networks just described is now rather well established, but as for other polysaccharide gel systems the rheological consequences of network formation have been less thoroughly explored. Smidsrød, however, has reported [205, 219] measurements of alginate gel rigidity as a function of polymer concentration, molecular weight, and M/G ratio. Systems rich in mannuronate, not unexpectedly, gave very weak gels of high turbidity, whilst gels rich in polyguluronate, were rigid and much less opaque. The guluronate gels were found to be very brittle, whilst gels containing an increased mannuronate content showed greater extensibility. Increased deformability in this case could be ascribed to the presence of increasing amounts of MM and MG sequences between junction zones.

Apart from M/G ratio, molecular weight and concentration were also found to influence gel rigidity in this study. For a typical alginate gel, other things being equal, the modulus was found to rise sharply with molecular weight until DP_n reached roughly 500, then it became almost constant and for a given molecular weight the modulus was reported by Smidsrod [219] to show a $G' \propto C^2$ dependence. He also found that divalent cation type had an influence on gel strength, the order of ability to produce strong gels by Group II cations being established as

$$Ba^{++} > Sr^{++} > Ca^{++} \gg Mg^{++}$$

Other rheological data have been reported for alginate gels, such as the account by Segeren et al. [220] of the results of dynamic viscoelasticity studies. In this work, the modulus of the gels tested was found to be independent of frequency in the range 10^{-1} to 10^1 Hz, and the ratio G″/G′ was found to be very small. G′ was found to increase linearly with temperature in the range 22 to 49 °C, and this fact, together with the previous information, led Segeren et al. to seek a description of the results within the framework of the classical theory of rubber elasticity. Accordingly, as others have done for carrageenan gels, they applied the well known equation $G' = CRT/M_c$ to data, and obtained various values for M_c, the molecular weight between cross-links, of chains assumed to show Gaussian statistical behaviour. The M_c values obtained were in reasonable agreement with corresponding results from swelling experiments, and this accord seemed to justify the rubber theory treatment in the alginate case. Observations from large deformation stress-strain experiments introduced an element of doubt, however, since linear viscoelastic behaviour ceased after a strain of 0.13, and the stress-strain curves steepened suggesting finite extensibility of the network chains. In the end, Segeren et al. concluded that the alginate gel network was a more complex structure than that envisaged by the rubber elasticity theory, and that the apparent applicability of the Gaussian coil description was probably accidental. A cancellation of opposing effects during deformation, such as the breaking of weaker bonds and the establishment of new stronger intersegmental contacts, seemed to be involved.

4.2.2 Plant Polysaccharides

4.2.2.1 Cellulose Derivatives

Cellulose itself, the commonest of plant polysaccharides, is a highly insoluble material which occurs in plant cell walls, and confers structural stability [221]. It is, of course, a principal component of timber and, in molecular terms, is a linear polymer of β-D-glucose residues 1:4 linked (Fig. 39).

Because of its extreme insolubility in the more usual aqueous media (e.g. salt solutions) cellulose itself is not a gelling biopolymer, but various kinds of cellulose

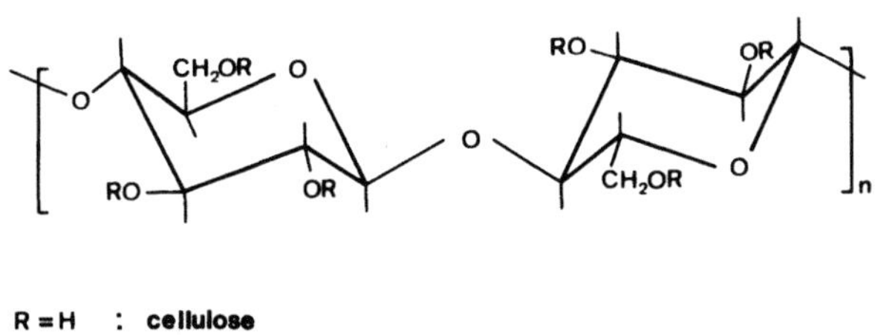

R = H : **cellulose**

$R = CH_3$: **O-methyl cellulose**

$R = CH_2CH(OH)CH_3$: **O-hydroxypropyl cellulose**

Fig. 39. Repeating structure of cellulose polymer based on 1,4-linked β-D-glucose residues. Structures of some important cellulose derivatives are also indicated

derivative can be made, and these form a variety of thickened solutions and gels. Thus, esterification with acids in the presence of a dehydrating agent, or by reaction with acid chlorides, or etherification by treating an alkaline cellulose solution with alkyl halide, produces a number of interesting cellulose derivatives capable of network formation in aqueous media. These will be discussed only briefly however, since once derivated they are no longer strictly biopolymers.

Hydroxypropyl cellulose (Fig. 39), for example, forms gels in water [222], although the mechanism appears not to be fully understood yet. In such a derivative (and indeed in most cellulose derivatives) [223] a balance exists between hydrophilicity and hydrophobicity, and the outcome of this balance in any particular set of conditions (temperature, solvent, etc.) determines whether the polymer is precipitated or associates in a more limited way to form a network. This picture can be further complicated, however, by the tendency of such materials to form lyotropic liquid crystals, because the polymer backbone is not very flexible, and so above a certain critical volume fraction (or concentration) an anisotropic phase can form [224, 225]. Conio et al. [226] have discussed this particular aspect of aqueous hydroxypropyl cellulose systems. An interesting discussion of the gelling properties of cellulose derivatives has also been given by Rees [172].

4.2.2.2 Pectins

Pectins [172] are cell-wall polysaccharides which, like cellulose, have a structural role in plants. They occur commonly as structural components of the cell walls of fruits, for example, and commercially, pectins play an important role as thickening and gelling agents in the manufacture of jams and jellies.

Although pectins are branched in their native form, as extracted, they are predominantly linear polymers of α-(1 → 4)-linked D-galacturonic acid residues interrupted occasionally (and perhaps systematically) by L-rhamnose residues (Fig. 40). In addition, as prepared, they contain varying degrees of methyl ester substituents, and are usually classed as low, or high, methoxy pectins. Preparations of low methyl ester content can be made to gel by controlled introduction of calcium ions, but gelation of both low and high methoxy pectins can occur without ions being involved at all.

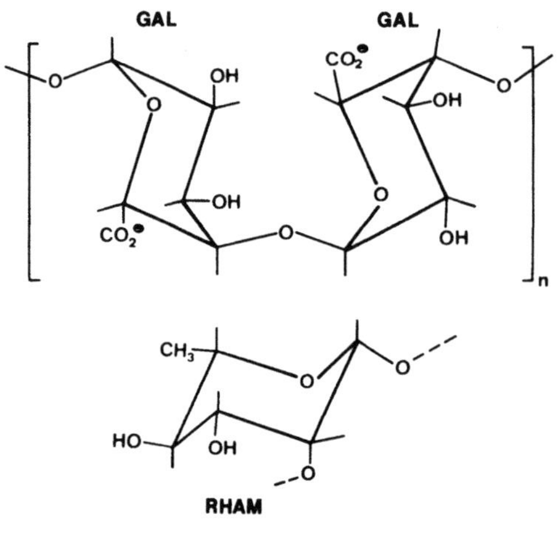

Fig. 40. Repeating structure of pectins based on 1,4- linked α-D-galacturonic acid residues. Linear sequence is occasionally broken by L-rhamnose residues also shown. Pectins usually occur in a partially esterified form (methyl ester)

This happens under conditions of low pH and decreased water activity, i.e. under conditions where intermolecular electrostatic repulsions are reduced. Morris et al. [227] have shown that under these conditions, the optimum degree of esterification for gelation to occur is around 70%. It appears that in the networks formed, cross-linking occurs via the formation of aggregates of chains of varying sizes. No definite aggregate stoichiometry is involved, and the sensitivity of the process to urea is believed to indicate non-covalent and non-electrostatic bonding forces (hydrogen bonds, hydrophobic bonds). These gels, made at low pH and at low water activity are sometimes thermoreversible, junction zones 'melting out' at fairly low temperatures (e.g. 40 °C) [228].

Pectin gelation at higher pH's, and in the presence of calcium ions, is reminiscent of the behaviour of the alginates, and indeed consideration of the structure of low methoxy pectin would encourage expectation of just such a parallel. The galacturonic acid blocks in pectins are almost mirror images of the guluronic acid blocks of the alginates, the configuration at one carbon only, being different. Thus, it would be anticipated that, as for the alginates, cross-linking of pectin molecules could occur by the 'egg box' model. This model for ion-induced pectin gelation has received support from techniques such as circular dichroism [229] and equilibrium dialysis [229], but interestingly, not from fibre diffraction [230, 231]. In the fibre the galacturonic acid sequences do not take up the buckled two-fold ribbon necessary for 'egg box' formation, but instead exist as three-fold helices.

The circular dichroism and equilibrium dialysis approaches, however, aimed at the fully hydrated state, provide clear evidence [229] for the two-fold ribbon conformation, the CD change on pectin gelation being identical to that occurring in alginate gel formation but opposite in sign because of the mirror image relationship. As for the alginate gels, the equilibrium dialysis approach indicates chain dimerisation rather than the formation of larger aggregates. Unlike for alginates, however, CD studies of films [232] formed from pectin gels reveal that a very substantial change in ellipticity occurs as the water content falls, and this has been interpreted as indicating a polymorphic transformation to the three-fold helix as the film state is approached. Additional evidence, reported by Powell et al. [233], involving experiments using galacturonic acid blocks derived from pectins, and using pectins which had been systematically de-esterified, have also confirmed the 'egg box' model for calcium-induced junction zones, and have suggested the need for sequences of at least fourteen galacturonic acid residues to be present if such zones are to form co-operatively.

No highly rigorous viscoelastic studies have been attempted for pectin gels, but for these materials at low water activity, and containing high methoxy pectins, compression test data are available and allow shear moduli and 'rupture strengths' to be estimated. In this work by Oakenfull [123, 124], a Saunders-Ward apparatus was also employed to measure the moduli of particularly weak systems, and shear modulus results were interpreted theoretically in terms of a junction zone model involving dimerisation of methyl esterified galacturonic acid segments. The driving force for association was described in terms of hydrogen bonding and hydrophobic interaction. A model relating shear modulus to concentration was constructed on the basis of Flory's theory of gelation, and the law of mass action applied to junction zone formation. This model, and its relationship to other theoretical approaches to the same problem, have already been discussed in Section 3.4. From the molecular point of view, it is of some interest that this treatment of pectin cross-linking assumes an exact dimersation of polymer chains whilst, as was mentioned above, other results by Morris and coworkers suggest a less stochiometrically precise situation. The discrepancy involved may ultimately be explained in terms of different materials and/or different gelation conditions.

4.2.2.3 Galactomannans

These polysaccharides [234] are derived from the endosperms of plant seeds, where they function as reserve materials utilised during germination. They are usually based on a β-D(1→4)-linked backbone of β-D-mannopyranosyl residues having single α-D-galactopyranosyl side chain stubs linked α-D-(1→6) (Fig. 41) and are used in the food industry as thickening agents and, in combination with other polysaccharides such as agarose and xanthan, as gelling agents.

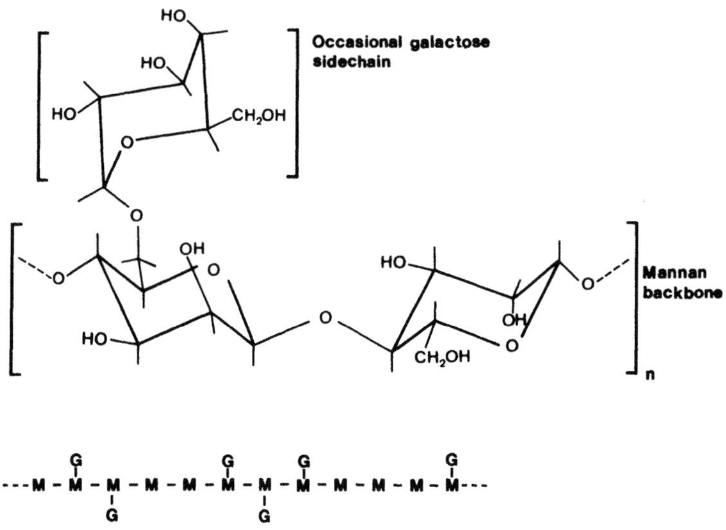

Fig. 41. Repeating structure of galactomannan backbone based on 1,4- linked β-D-mannopyranosyl residues. Note occasional sidechain substitution involving 1,6- linked α-D-galactopyranosyl residues

As extracted from various seed sources, galactomannans can differ widely in their galactose/mannose composition ratios. At a given ratio, however, it turns out that they can also differ in terms of the distribution of side-chains along the mannan backbone. This so-called fine structural aspect of galactomannans can influence the viscoelastic properties of galactomannan solutions though, in fact, the degree of side chain substitution is the principal determining feature. It was once believed that galactomannans contained blocks [235] of galactose side chains, i.e. that they were block co-polymers, but more recent studies [236] based on enzymic degradation and computer modelling suggest highly random (though, in general, not totally random) distributions of galactose stubs (Fig. 41).

Since the isolated mannan backbone chain is water-insoluble, it might be supposed that substituted galactomannans would have the capacity to form gel networks, as blocks of mannose residues are available to form junction zones even in the randomly substituted polymers. In practice, however, solutions of galactomannans, though viscoelastic [237, 238], do not form strong gels, though freeze-thaw treatment can lead to precipitation, the least substituted polymers precipitating most readily. It appears that, in general, galactomannans in solution tend to form entanglement networks (see Sect. 4.3 below) rather than become cross-linked by specific inter-chain associations.

The properties of the galactomannans, including their propensity for interaction with other kinds of polysaccharide have been extensively reviewed by Dea and Morrison [234].

4.2.2.4 Starch

Starch is a complex polysaccharide substance of enormous importance in human and animal nutrition. It occurs in many forms, e.g. potato starch, corn starch, tapioca starch, rice starch etc., and in plants it functions as a reserve polysaccharide.

As a raw material extracted from a plant source (flour obtained by milling, for example) starch normally has a granular structure, (particles in the size range ~2–100 μ) the granules themselves involving a complex packing arrangement [239] of starch polymers, and containing other bio-molecular components such as lipids and proteins. The starch fraction itself consists of two classes of starch molecule, one branched and called amylopectin, the other linear, and called amylose. These are based on repeated glucose residues — amylose being an α(1→4)-linked-D-glucose polymer whilst amylopectin also involves D-glucose residues but has both α(1→4) and α(1→6) linkages (Fig. 42). Though some exceptional starches occur which have unusually low or high amylose contents, amylose is normally the minor constituent by weight, accounting for only ~25% of the total starch, and in the granule amylose and amylopectin are packed together in a way that is not yet fully understood. That the granules are partly crystalline (A, B and mixed AB forms) has long been known

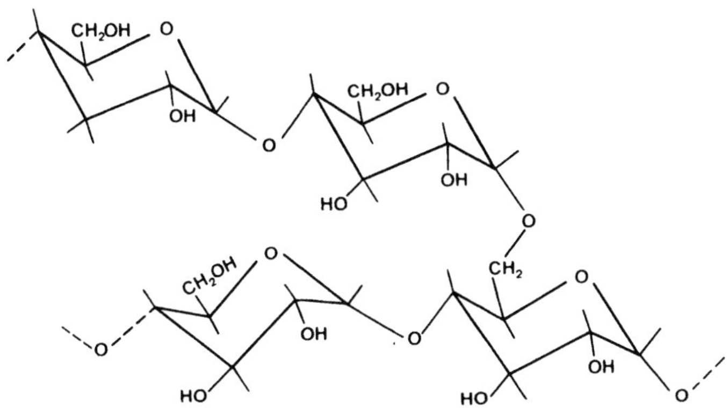

Fig. 42. Branched structure of the amylopectin component of starch based on 1,4- and 1,6- linked α-D-glucose residues. Note that amylose, the other polysaccharide component of starch, involves only the 1,4 linkages and is a linear polymer

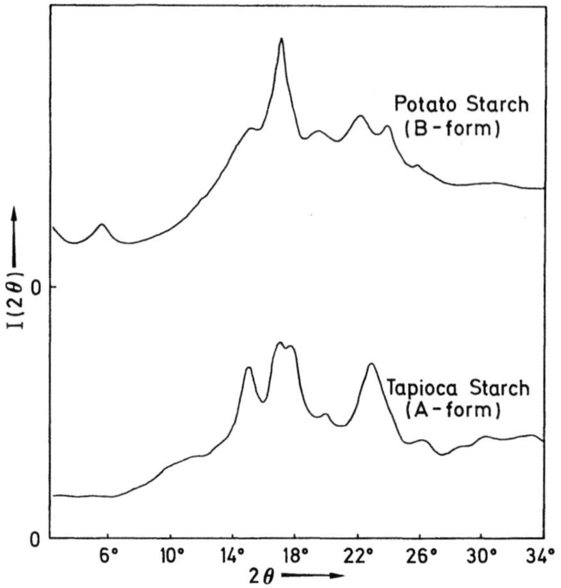

Fig. 43. Wide-angle X-ray diffraction data for two starches (potato and tapioca) in powdered granular form. The amylopectin fractions crystallise as distinguishable A and B polymorphs

and is easily demonstrated by wide-angle X-ray diffraction experiments (Fig. 43), the crystalline fraction being believed to be composed largely of amylopectin molecules. The molecular weight distributions of both the amylose and amylopectin fractions are broad and individual molecular weights can run into millions. When heated in water above a specific temperature known as the gelatinisation temperature starch granules in suspension swell suddenly, and irreversibly, and the amylose component leaches into solution. If the starch concentration is high enough this mixture of swollen granules and free polymer behaves as a viscoelastic paste. On cooling, thickening occurs, opacity increases, and an elastic gel is formed. Ring and Stainsby[240] have described the resulting gels as composite materials consisting of amylose gel matrices 'filled' with swollen granules.

The aggregation and gelation characteristics of both the amylose and amylopectin fractions of starch have received attention in recent publications. Miles et al[241], for example, have followed network formation by linear amylose chains using a combination of turbidity, X-ray diffraction, and shear modulus measurements. In this work hot amylose solutions (7% w/w) were quenched from 90 to 32 °C, and it was shown that both the shear modulus and the turbidity increased rapidly during the first minutes of the resulting aggregation process. On the other hand the development of crystallinity, as monitored by wide-angle X-ray diffraction, was slow and was still incomplete after many hours. It appears from this, and from previous studies of amylose gelation, that as hot solutions of the polymer are cooled, rapid network formation occurs through the formation of ordered junction [242] zones based on amylose double helices. The association process leads to turbid gels so it is likely that multiple aggregations of helical chain segments are involved and certain that the polymer network is phase separated on a micro scale into polymer-rich and polymer deficient regions. Subsequently, the aggregates of helices become bigger and develop into crystallites capable of X-ray diffraction (B-form). It is perhaps surprising that the shear modulus is unaffected by this last feature of the network building process, but the measurement frequency of the pulse shearometer employed by Miles and coworkers is high (~200 Hz), and a different result might have emerged at lower measurement frequencies (such as 1 Hz).

More conventional rheological studies of amylose gels have also been performed. Ellis and Ring [74] for example, have measured shear moduli using a Saunders-Ward gelometer and have related these to amylose concentration. A critical concentration C_0 of approximately 1.0% w/w was indicated by this work as well as a C^7 power law dependence of modulus on concentration in the range ~1.5% w/w to ~7.0%, and although Ellis and Ring have contrasted this exponent with the value of ~2.0 more commonly found in the biopolymer gel literature, the concentration range addressed by their experiments, when re-written as C/C_0, covers sufficiently low values of this reduced variable, that the C^2 part of the master curve discussed in Section 3.4 is unlikely to have been accessed. At concentrations corresponding to C/C_0 values less than 10, a much higher power law dependence of modulus on concentration than C^2 would be expected. In their rheological studies, Ellis and Ring have also remarked upon the high elasticity of amylose gels, and upon the tendency for the shear modulus to fall slightly with temperature in the range up to 100 °C. Amylose gels have a high melting point (in excess of 100 °C) and hence show pronounced hysteresis. They are also radically influenced by thermal history and are clearly remote from true thermodynamic equilibrium. In Ref. [74] it was concluded that amylose gels, though highly elastic, were unlikely to be describable within the framework of the classical (entropic) theory of rubber elasticity.

Structural and rheological data are also available for the other polysaccharide component of starch, amylopectin. Using a starch extract containing mainly this branched material, Ring has shown [243] that it aggregates very slowly when solutions are cooled and that, eventually, after long periods, there is a substantial re-establishment of crystallinity. The gels formed are highly turbid as in the amylose case, but they can be melted below 100 °C, and hence the cross-linking and crystallisation processes are reversible. It is suggested that the high molecular weight, and highly branched, amylopectin species slowly associate, with a concomitant crystallisation of short branches situated near their surfaces. Although this process probably also takes place during gelling of whole starches containing both (branched and unbranched) components, in the whole starch situation the amylopectin remains trapped in granular form after gelatinisation, and the gel which results is based on an amylose network in which the swollen granules are suspended. The mechanical properties of such a gel are different from those of the pure amylose and amylopectin gel systems, and for starch gels a linear dependence of modulus on concentration has been reported. This may be a consequence of the composite nature of the starch gel system, but could on the other hand reflect on inadequate C/C_0 range of measurement, making it impossible to establish the functional form of the modulus-concentration relationship. Clearly, further studies of the structures and rheologies of gels based on starch and its components are required.

4.2.3 Microbial Polysaccharides

Most of the polysaccharides discussed so far have had structural significance for the plants and seaweeds containing them, and have occurred predominantly in cell walls. Polysaccharides of scientific and commercial interest also occur outside the cells of certain bacteria and fungi (yeasts and moulds) and may be covalently attached to the cell wall or secreted unattached into the growth medium. These are the microbial exopolysaccharides [244] and there are a great variety of types known, some of which find extensive industrial use.

As yet, unlike in the plant situation, there is no clear understanding of the biological function of the microbial exocellular polysaccharides, though some it seems may be involved in recognition exercises [245] enabling bacteria (pathogens) to become attached to plant cell walls (hosts). This appears to be true of xanthan, a well-known bacterial polysaccharide from the organism *Xanthomonas campestris*, which acts as a plant blight.

In molecular terms, the microbial polysaccharides show great diversity of covalent structure [244], and are generally much more complicated in this respect than the algal and plant polymers already discussed. Thus, they may be branched as in the case of dextran, or based on a more complex repeat unit than a simple disaccharide. They may also take the form of a linear backbone, carrying complex side chains, as in the xanthan situation where every second residue of the β-1,4-linked glucose backbone carries a trisaccharide side chain (Fig. 44).

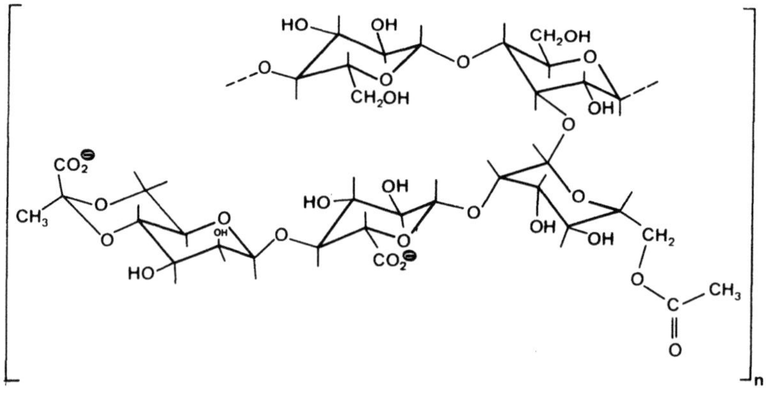

Fig. 44. Xanthan structure based on a 1,4 linked β-D-glucose backbone and complex trisaccharide sidechain occurring every second glucose residue

In some respects, however, particularly in rheological terms, bacterial polysaccharides resemble the plant and algal polysaccharides in behaviour. Conformational disorder-to-order transition occur [246] under conditions of changing electrolyte concentration, temperature etc., and these are accompanied by rheological changes including the formation of strong gels. As in the previous plant and algal examples, junction zone formation, through association of ordered regions, has been proposed [247]. It seems, however, that in some cases at least, the junction zones are much less strong [248, 249] than those in, for example, alginates, and the networks formed are essentially transient. In this case the corresponding gels are described as weak and show pseudo plastic behaviour. The xanthan polysaccharide mentioned above is an example of a bacterial polysaccharide behaving in this way and its rheology will be discussed in more detail in Section 4.3. devoted to entanglement networks and weak gels.

Bacterial polysaccharides can, however, form strong gels of the alginate and carrageenan type, articles having been published describing this behaviour for the materials curdlan [244], gellan [250] and XM6 [247]. Although it is believed that association of ordered junction zones (sometimes with ion involvement) is the origin of this behaviour, studies of these systems are still in progress and their detailed structural and rheological properties remain to be established.

4.2.4 Animal Polysaccharides

Because of the vastly different structures and motilities of animals, as opposed to plants and bacteria, it might be supposed that animal organisms would make much less use of polysaccharide macromolecules, and that most polysaccharides worth investigating would derive from plant and bacterial sources. However whilst it is true that very great differences exist between plant and animal polysaccharides, and between the roles of polysaccharides in these different types of organism, animals rely heavily on polysaccharides for their successful functioning, and the animal polysaccharides have amongst others, both structural and storage tasks to perform. They are also required to lubricate joints and various passages within animal structures such as the alimentary canal, air passages etc. In general, it may be said, that whilst in plants polysaccharides are called upon to supply rigid, elastic, structures; in animals viscoelastic behaviour is more important, and in consequence polysaccharides in animals, either alone, or in combination with other types of biopolymer, supply necessary viscoelastic properties. Two examples of classes of animal polysaccharides fulfilling this function are now considered.

4.2.4.1 Proteoglycans

An example of a combination of protein and polysaccharide components promoting viscoelastic character is the type of macromolecule known as a proteoglycan [251]. Proteoglycans occur in many parts of animals, and in a variety of tissues, but they are predominantly associated with connective tissues, such as cartilage. Although proteoglycans are somewhat complex and polydisperse species, they have a general structure involving a protein backbone to which is covalently bound many linear polysaccharides known as glycosaminoglycans (Fig. 45). These are polyanionic polysaccharides of variable chain length, and consist of repeating disaccharide units, each of which contains a nitrogen-bearing hexosamine sugar, and each of which usually carries negatively charged sulphate ester, or carboxylate groups. The glycosaminoglycans found in cartilage go under such names as chondroitin sulphate, keratan sulphate, and hyaluronic acid, but other kinds occur in other tissues. The hyaluronic acid polymer [34, 252–254], whose repeating disaccharide unit appears in Fig. 46, forms a polymeric backbone to which a large number of proteoglycan molecules attach (Fig. 45), the whole complex aggregate being able to co-exist with a fibrous collagen matrix [34, 251]. The resulting composite aqueous gel forms an extracellular matrix for cartilage cells, and it is believed that the proteoglycan role in this gel, is to retain water and to supply appropriate viscoelasticity.

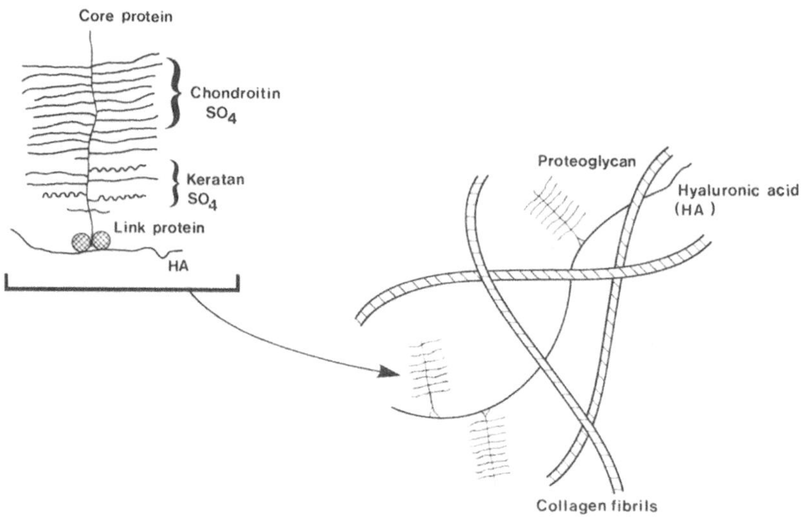

Fig. 45. Schematic representation of proteoglycan structure and indication of its role in network formation with hyaluronic acid and collagen

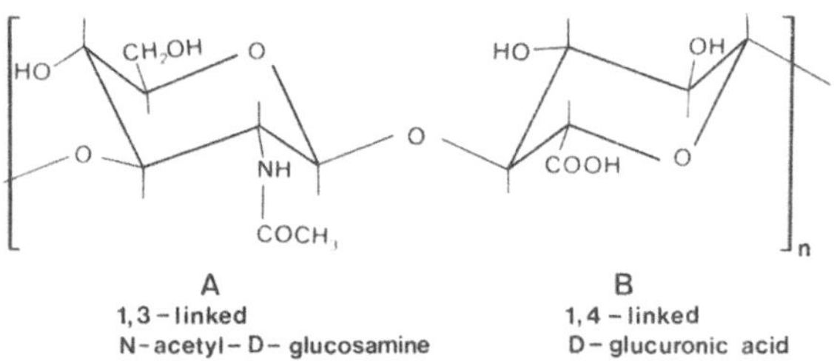

Fig. 46. Repeat structure of hyaluronic acid polymer

Whilst the properties of whole cartilage are very complex, glycosaminoglycans can be separated from proteoglycans, and great interest has attached to their rheological properties in aqueous solution. Hyaluronic acid, for example, has been extensively studied [26, 252–254] in this respect both by structural and by rheological techniques and in general, strong gels are not formed. Hyaluronic acid will be mentioned again in Sect. 4.3 below on transient networks. Meanwhile polymers closely related to proteoglycans and involved in lubricative function, the mucus glycoproteins, will be discussed.

4.2.4.2 Glycoproteins

In animals, one of the major roles of biopolymer gels is as mucus, lining the respiratory and intestinal tracts [255,256]. For example, in stomach and duodenum, the adherent layer of mucus gel has been proposed to protect the underlying cells (mucosa) from acid and pepsin (proteolytic enzyme) in gastric juice and from mechanical damage by compressional and shear strains during digestion.

The major molecular constituents are biopolymers known as glycoproteins which contain both protein and carbohydrate moieties linked covalently. The glycoprotein of gastric mucus, for example, contains over 70% by weight of saccharide, typically arranged as 500–800 short chains (say 2–20 saccharides) situated around a central protein core. Since the saccharide units are themselves usually charged, and the side chains are short and stiff, an analogy is often made between this structure and that of a bottle brush, where the carbohydrate chains represent the bristles, and the protein core represents the central wire support [257]. The mucus, itself, then consists of a number of these 'bottle brushes' attached by interchain disulphide bridges to a small central protein unit; the 'windmill model' [258]. Evidence for these levels of structure are based mainly upon molecular weight (for example gradient ultracentrifugation) and intrinsic viscosity measurements [259]. By treating the separated mucus glycoprotein with various enzymes and reagents which break down the macromolecule to different extents, the structural complexity may be established.

The mucus itself is secreted by mucosal cells as microgel particles (dimensions say 10 μm in diameter), which then form a macroscopic layer of thickness say 5–200 μm, over the tissue walls. Although many studies have been made, of an essentially biochemical nature; because of the clear link between biological function, structure and rheological properties, a number of groups have made measurements of the mucus gel itself.

Litt and co-workers have mainly concentrated upon bronchial mucus, using a variety of instrumentation [260, 261]. For example they have exploited a creep rheometer to measure J(t) vs. time for a number of mucus systems, and used the technique to monitor the efficacy of different treatments for 'clearance' of inflammation of the chest and respiratory systems. A great deal of this work involves extractions of mucus and *ex vivo* treatment with various trial formulations.

A similar approach is also seen for gastric mucus, where physiological interest in ulcer formation has suggested that high concentrations of bile salts (essentially surfactant systems) occur in patients with ulcerations. Martin et al. [262] have treated mucus extractions with bile salts and shown that this results in a decrease in reciprocal compliance.

Allen and co-workers [263–265] have recently published a series of papers involving a detailed rheological description of pig gastric mucus, probably the best studied system from the structural viewpoint. Dilute solution viscometry of both native and extracted and purified glycoprotein show that the glycoprotein itself is highly expanded and solvated ($[\eta] \sim 320 \text{ ml g}^{-1}$ in 0.2M KCl), but approximately 'spherical' — consistent with the structural evidence above. The concentration dependence of viscosity for the purified glycoprotein is consistent with simple polymer entanglement, but at higher concentrations the plateau modulus extends over the entire accessible frequency range (Rheometrics instrument). Mechanical spectra for the native gel are almost frequency independent and strain independent; however on shearing the gel complete recovery takes place only after ~2 hours.

At least some of the properties observed for gastric mucus are consistent with entanglement of branched chain polymers, but others, notably the presence of obvious sol and gel fractions *in vivo*, are not. Furthermore, the gels 'reconstituted' from purified glycoprotein, and at the same nominal polymer concentration, were appreciably more strain sensitive than the 'native' material. Bell et al. [263] have suggested that *in vivo* the branched carbohydrate side chains of adjacent macromolecules interdigitate, giving rise to relatively stable and long lived intermolecular associations, largely (though perhaps not exclusively) topological in nature. Results from Sellers [265] tend to confirm this, because data from glycoproteins with different numbers of carbohydrate residues in the side arms show a quite clear gradation in properties from those consistent with simple entanglement through to 'strong' gels, as the number of side arm residues is increased.

4.3 Polysaccharides Forming Entanglement Networks and Weak Gels

4.3.1 Introduction

Almost all of the preceding discussion in this Section has covered 'strong gels', defined in terms of a characteristic resistance to strain, and stress/strain profile up to failure. It is usually considered that there are qualitative differences between these systems and the 'entanglement networks' induced by increasing chain length/concentration of linear polymer in a solution. These differences may be summarised as

1) On dilution, the strong gel swells, whereas the entanglement network passes back through the C* transition and behaves as a dilute solution.
2) At low frequencies the former still behaves as a solid, whilst the latter flows as a liquid (i.e. the former has an infinite relaxation time, as well as the usual Rouse-Doi-Edwards relaxation spectrum for disentanglement).
3) The former ruptures, and fails complety as the strain is increased above the ultimate strain, and subsequently it has no macroscopic integrity.

Particularly for biopolymer systems, where non-covalent bonding is the rule rather than the exception, a number of systems exist which, even well above C_0, behave in an intermediate fashion and show some of the properties discussed by Ferry and co-workers [266] as arising from hyperentanglements, i.e. specific molecular interactions occurring over and above those due purely to physical entanglements. Here these are designated "weak gels", cf. Ref. [76] and again the distinction should be noted between the definitions of "weak gels" adopted by de Gennes [41], van den Tempel [267] and ourselves. The van den Tempel definition is purely phenomenological, and is particularly concerned with the extreme strain sensitivity of colloidal network systems, i.e. Bingham plastic materials. The distinction, is in truth, unnecessarily confusing, but reflects the looseness of language which abounds in this area.

To illustrate the differences between the strong gels, weak gels and entanglement network systems it is useful to reproduce Figs. 47 and 48 [76]. The former shows mechanical spectra for a strong gel (gelatin), a weak gel (xanthan) and an entanglement network (guar galactomannan) cf. Fig. 14. All of these are measured at different nominal concentrations, although the degrees of space occupancy, in terms of the C/C* (or C[η] parameter) were quite similar. Nevertheless, the spectra for agarose and for xanthan are similar, showing little frequency dependence over the experimental timescale. The result for the guar is much more like that expected for an entanglement system [238]. Increasing the concentration to achieve the same space occupancy as xanthan, does lower the frequency at which a plateau modulus is seen, but conversely, diluting the xanthan, does not greatly affect the spectrum of this system until much lower concentrations are reached. Fig. 48 [268] sketches the strain dependence

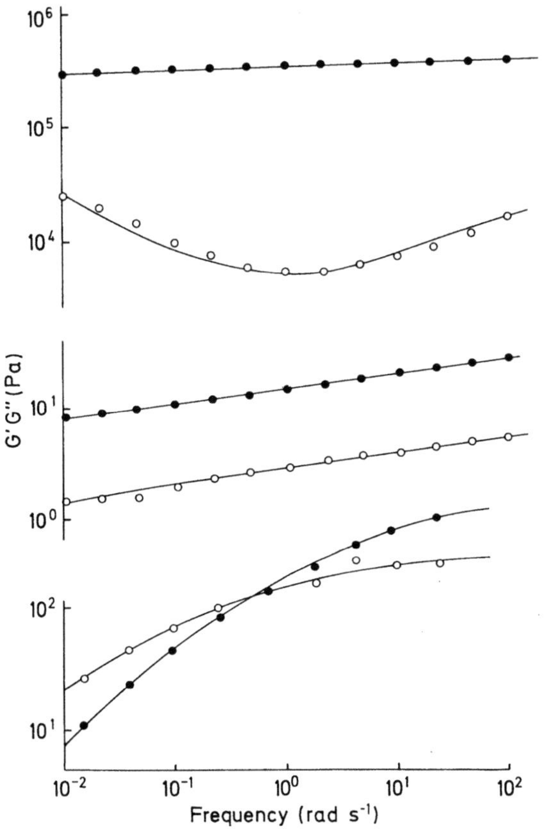

Fig. 47. Typical mechanical spectra (G′ filled circles, and G″ open circles) for gel (top), weak gel (centre) and entanglement network systems, plotted against frequency. Data are for 25% gelatin (Ref. [96]), 0.5% xanthan (Ref. [248]), and 3% guar (Ref. [238]). The degree of volume occupancy (c[η]) is approximately the same for all three systems

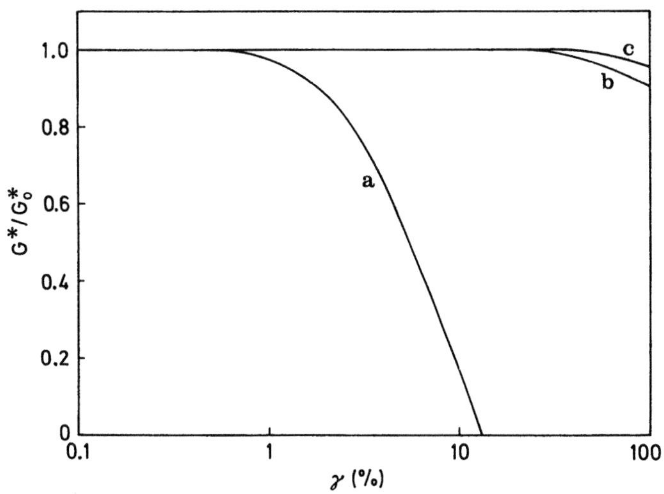

Fig. 48. Strain dependence of shear modulus for typical (a) "weak gel", (b) entanglement network and (c) gel systems. Reproduced with Permission from "Biophysical Methods in Food Research", edited by H. W.-S. Chan, Fig. 14, p. 167, Blackwell Scientific, 1984. Copyright 1984 Society of Chemical Industry

(in terms of the reduced modulus G/G_0) of these three systems. Here xanthan "gel" is much more strain dependent than the strong gel or the entanglement network.

The major distinction between the entanglement network and the "weak gel" may be made, however, by a steady shear rate experiment (clearly this cannot be performed at all for the true gel which will rupture rather than flow). Here the terminal slope in a plot of $\log\eta$ vs. $\log\dot{\gamma}$ will be ~ 0.72 for the entanglement network [9], but ~ 0.9 for the "weak gel" [269]. Such high exponents are sometimes seen for rigid rod or polymer liquid crystal systems [270], so some discussion of the molecular aspects of rheology in such systems is appropriate at this stage.

4.3.2 Molecular Rheology of Xanthan Polysaccharide

Xanthan is an anionic polysaccharide produced commercially by fermentation of the bacterium *Xanthomonas campestris*. Its primary structure (Fig. 44) is essentially that of a cellulose backbone, but with charged trisaccharide side chains attached to alternate residues of the polymer backbone to give a pentasaccharide repeating sequence [271]. The majority of studies of xanthan in the literature have been "molecular" (including optical rotation, NMR, d.s.c., circular dichroism) [246,272–274], but the interpretation of some of these have been complicated by the nature of the macromolecule. Since xanthan is a polyelectrolyte its chain dimensions would be expected to change in response to changes in ionic strength. Under conditions of elevated temperature, and/or comparatively low ionic strength, this is indeed the case, and the local chain flexibility appears similar to other coil-like cellulose derivatives. On cooling, or on addition of salt, however, the molecule undergoes a cooperative conformation transition to a rigid ordered structure [274].

The exact form of this ordered structure is still somewhat controversial. The order-disorder transition has first-order kinetics, is fully reversible, and shows no thermal hysteresis. From this evidence Morris and co-workers [246,274] suggested a single helix stabilized intramolecularly by ordered packing of side chains along the polymer backbone. However, the recent work by Sato et al. [275,276] appears to establish unequivocally (mass per unit length) that a dimeric helical form occurs with a persistence length of ~ 120 nm and the hydrodynamics of a Yamakawa-Fujii wormlike chain. This confirms the earlier light scattering results of Paradossi and Brant [277] and finds agreement in more recent results by Coviello and coworkers [278]. There need be no contradiction between these apparently disparate views if the dimeric species consists of paired single helices (as distinct from a wound double helix). For derivatised xanthans different results are apparently obtained, and Muller et al. [279] have demonstrated that a fully pyruvated xanthan sample exists as a single helix. Confirmatory results have also been found by Coviello and co-workers [278] on partially derivatised samples.

As far as the rheological aspects of xanthan behaviour are concerned, such molecular details are currently only of indirect significance, because the macromolecular characterisation is greatly complicated by the presence of a microgel fraction, and by the number and species of cations present and the ionic strength. What would appear unequivocal, however, is that the "weak gel" rheological response involves the shear breakdown of a supramolecular aggregate [248,280]. The fast recovery of this structure implies the reformation of non-covalent intermolecular bonds. Since the ordered form of the macromolecule is so stiff [281], liquid crystal structuring has been suggested as crucial to the weak gel formation. However, although large stress birefringence is seen at high concentrations, and a cholesteric form is observed [282], the "weak gel rheology" occurs at concentrations more than 10X lower than the phase volume producing anisotropic behaviour. Combined rheo-optical [249] and viscoelastic measurements for xanthans treated with urea, especially for single cation exchanged samples [280], would seem to confirm that intermolecular ion and hydrogen bonds are involved since under appropriate conditions the weak gel properties may be eliminated although the xanthan macromolecule is still in the ordered, worm-like form.

It is of interest that similar effects are observed for the glycosaminoglycan hyaluronic acid (Sect. 4.2.4), which occurs as the intercartilaginous lubricant in, for example, human knee joints. At acid pH (2.5) and .15M salt (ionic strength) weak gel "putties" are formed [283]. Other conditions of pH and ionic strength produce mechanical spectra much more akin to entanglement networks [252,253]. This effect is much more pronounced than can be attributed to the change in macromolecular dimensions expected to accompany the above changes in conditions.

5 Networks from Globular and Rod-Like Biopolymers

5.1 Introduction

In the previous Section network-forming biopolymer systems were described in which the aggregating species in the sol state could be regarded as polymer chains with inter-residue bond angles statistically distributed over a wide range of conformational states. Organisation of these macromolecules into physically cross-linked networks involved generation of disorder-to-order transitions by changing temperature or introducing appropriate electrolytes. The process of network formation could be seen as a limited form of polymer crystallisation leading to very many microcrystalline junction zones connecting individual polymer chains. In these networks the physically formed junction zones replace the point (covalent) cross-links of vulcanised rubbers and, of course, unlike in the rubber example, a solvent (water) is present in large excess. As the Introduction to this article has suggested, such networks are closely analogous to those underlying certain types of synthetic polymer gel such as those formed from isotactic polystyrene solutions, and it is for this reason that they have been described first, in preference to accounts of other perhaps less familiar types of gelling biopolymer system.

Biopolymers in solution, however, particularly proteins with some specific biological function, often occur as compact particles with the constituent polymer chain (or chains) wound into some specific conformation. These particulate biopolymers can assume globular, or rod-like, forms and under certain circumstances they can be induced to associate and form gel networks. Where the natural function of the biopolymer is related to such an aggregation process (see, for example, the actin and tubulin systems discussed below) highly organised rod-like structures can arise even from globular monomers and in this situation there is no guarantee that further strong cross-linking of these species will occur. Instead, they may behave in solution purely as rods (see Sect. 3.7) and form entanglement networks, or at higher concentrations, liquid crystalline phases and under certain conditions of concentration and solvent quality they may even collapse into the intrinsically unstable microphase-separated networks described by Miller for helical polypeptides in certain solvents (see Sect. 5.5.1 below). Where the rods have some flexibility, however and can intertwine, or associate laterally in a less total way than is implied in the formation of Miller's gels, more conventional gel networks may arise, such as those underlying the blood gels which form when fibrin polymerises (Sect. 5.2.5, below). The association (induced, for example, by thermal or chemical means) of other types of compact biopolymer, such as many globular proteins, whose natural function (e.g. enzyme function) has no aggregation requirement, can also occur and be much less organised still, involving aggregation of the monomers with extensive branching (Sect. 5.3) and clumping. The distinction between this last situation and the fibrin case is not a sharp one, however, as under certain solvent conditions, gel networks may assemble from denatured globular proteins by cross-linking of predominantly linear aggregates.

In the present Section, gel formation from particulate protein systems is discussed at length, and examples classed in terms of whether the initial monomer can be regarded as globular or rod-like. The division produced is not highly significant, however, as fibrin, the monomer of the fibrin network, is probably a short rod in the form in which it enters the gel structure (Sect. 5.2.5) and myosin, the principal protein in muscle (Sect. 5.5.2), which is correctly categorised as rod-like, often shows aggregation behaviour uncharacteristic of rods and similar to that arising when globular protein solutions are heated (Sect. 5.3.1).

The classification attempted below is therefore somewhat arbitrary, and overlapping behaviour between systems will undoubtedly be apparent. In terms of monomeric structure, however, the gelling casein system (Sect. 5.4) is unique for whilst this undoubtedly contains globular particles, these are aggregates of smaller spherical submicelles which themselves contain many protein monomers in rather disorganised conformational states. Nevertheless, the association of casein micelles to form large aggregates and gel networks has many features in common with the other systems discussed, and is justifiably included amongst them. Modes of inducing compact protein particles to associate include, changes of solvent quality (ions, cosolutes, pH, temperature), heating to produce thermal unfolding, addition of specific cross-linking agents (for example, in blood agglutination and antibody-antigen systems), modification of the particle surface by enzyme reaction (casein-rennet system), chemical denaturation, or some combination of these.

In network formation of the type discussed in this Section, where some particulate identity is maintained by the aggregating species (quite common for protein particles) the gels formed can be thought of as more closely related to networks of colloidal particles than to the systems discussed previously in Section 4. Even for casein gels, however, for which the colloidal description has often been advanced, the particle-particle interactions involved are significantly more directional and complex than is envisaged by conventional treatments (DLVO theory) of colloidal stability.

5.2 Globular Proteins Forming Ordered Fibrous Assemblies

5.2.1 Actin

5.2.1.1 Introduction

The thick and thin filaments forming muscle myofibrils will be discussed later (Sect. 5.5.2) in the context of the aggregation and gelling properties of the muscle protein myosin. The globular protein actin is also present in muscle where it is a principal component of the thin filaments [284]. There it occurs in conjunction with a thread-like protein tropomyosin, and troponin, another globular protein. Actin plays a principal role in muscle contraction as it is the interaction between actin and myosin which is at the heart of the mechanism of muscle function.

In addition to its importance in muscle, actin is an intrinsic component of mechanisms of cell motility in general, for it is part of the cytoskeleton of cells [285], where it occurs with other proteinaceous substances, such as tubulin, and tubulin associated proteins. The cytoskeleton filaments which result have important cellular functions related to structure, cell movement and cell shape change, and clearly as a component, actin, in its fibrous form, is an extremely important structural protein. In this Section its aggregation and network-forming capacity both alone, and in conjunction with other molecular species, are considered.

5.2.1.2 Structures of G and F Actin

G-actin, the globular subunit of actin polymers, has a molecular weight of approximately 45,000 and a diameter of roughly 5.5 nm [284]. It can be extracted from skeletal muscle, and purified samples can be induced to form long rod-like aggregates (F-actin) on addition of salts (0.1M KCl, for example). The filaments formed are very similar in organisation to the actin chains in natural muscle, and over twenty years

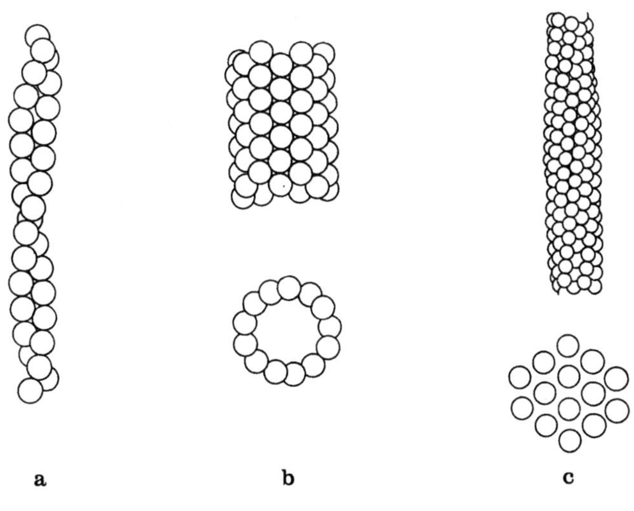

a b c

Fig. 49a–c. Organisation of protein subunits to form rod-like aggregates of (**a**) actin: (**b**) tubulin) (**c**) haemoglobin-S

ago Hanson and Lowy [286] showed, using electron microscopy, that the filaments were composed of two helical chains, or strands, of actin monomers wound round a common axis (Fig. 49). The filament diameter was found to be approximately 8.0 nm. Whilst in muscle, the thin filaments are well defined species, and are comparatively short, polymerisation *in vitro*, it seems, produces a distribution of lengths, some very long polymers being present.

5.2.1.3 Rheological Properties

Actin polymers in aqueous solution have been extensively studied rheologically. The "thixotropic" character of actin solutions has long been known, and an early quantitative study by Kasai et al. [287] in 1960, pursued this aspect via measurements of viscosities and rigidities of F-actin preparations. A transient network was inferred from the results.

Some years later Maruyama et al. [288] carried out a similar investigation. They believed that the samples used in the Kasai study could have contained impurities (α-actinin and tropomyosin) and that these could have introduced cross-links into F-actin preparations. Studies using steady shear viscosity were performed on highly purified F-actin solutions at pH 7.2 and at various concentrations, and these measurements indicated strong shear-thinning behaviour above a critical concentration of 0.015 mg/ml. Below this, the F-actin depolymerised. The viscosity at low shear rates (0.0005 s^{-1}) was extremely high, and in general it was found that the product of viscosity and shear rate was constant, i.e. the system possessed a finite yield stress.

Measurements in oscillatory shear, also by Maruyama and co-workers, allowed G' and G'' to be established, and at the concentrations studied (1.2 mg/ml) these were very small indeed (1.0 to 1.5 dyne/cm^2). The frequency of measurement extended from 1/60 Hz to 1/3 Hz, and the growth of G' and G'' as functions of time during actin polymerisation was followed at a particular frequency until equilibrium was established ($\sim$30 min). Plots of equilibrium values for G' against concentration, in the range 0.5 to 3.0 mg/ml, suggested a $C^{2.5}$ dependence. At all frequencies G' was greater than G'', the ratio G''/G' being roughly 0.5, and both values tended to increase slightly with frequency. The ready fall in shear modulus with increased strain which was also recorded, and the viscosity behaviour referred to above, led Maruyama et al. to conclude that their actin system was a temporary network easily destroyed by external force. It was not clear, however, whether the temporary cross-links could be ascribed to entanglement, or to non-covalent interactions, or indeed to residual impurities still present in samples.

Subsequent rheological studies of F-actin have led to similar conclusions. Jen et al. [68], for example, used a Rheometrics Fluids Rheometer to make sensitive measurements of the shear modulus of F-actin solutions. Again, very small modulus values were recorded, and a temporary network involving weak forces of association was diagnosed.

Zaner and Stossel [289], in their work on actin viscoelasticity, concluded that a temporary network was involved, but they favoured an entanglement description, and showed that modulus-frequency data agreed well with Jain and Cohen's theoretical predictions [290] for a solution of rigid rods (Fig. 50). This description also found support from dynamic light scattering studies by Carlson et al. [291], and Fujime and

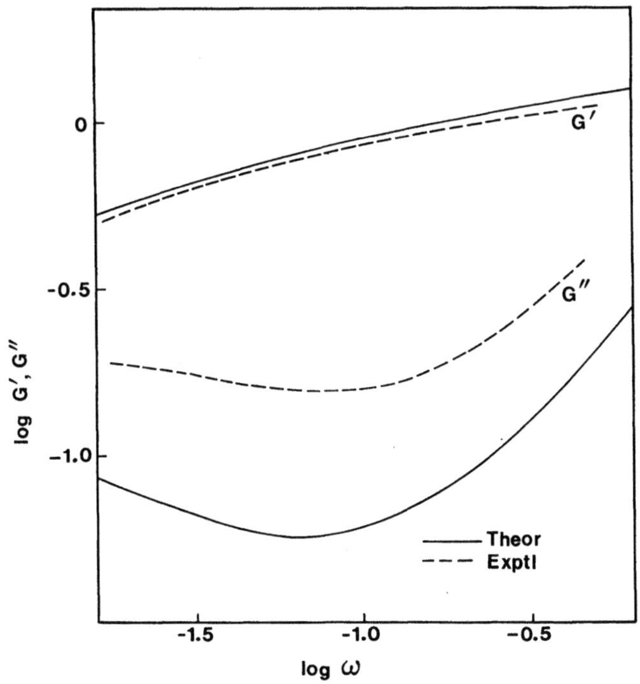

Fig. 50. G′,G″-frequency data for F-actin solution (1 mg/ml). Weight average actin filament length was 2.5 micron. Theoretical predictions for solution of rigid rods based on Jain and Cohen's theory appear as solid lines. Reproduced with Permission from J. Biol. Chem. *258*, 11004, (1983), Fig. 7, p. 10007

Ishiwata [292,293], but these authors disagreed about the degree of flexibility of the F-actin polymers. Interestingly, Carlson et al. were able to demonstrate a loss of freely diffusing character in F-actin rods, as myosin head-group fragments (Sect. 5.5.2) were added. The myosin, not unexpectedly, had a cross-linking effect, and a transition to a more constrained network was observed.

5.2.2 Tubulin

5.2.2.1 Introduction

Microtubules, derived from the globular protein tubulin, are highly organised hollow rod-like assemblies, and are commonly found in the cytoplasmic material of cells [285]. There, they can be assembled and disassembled by processes involving associated proteins, which influence the nucleation step which is part of the co-operative assembly process [294,295]. Like F-actin, the microtubules are part of the filamentous structural components of cell cytoplasm, and are involved in several processes including mitosis. The microtubules are large structures, a recent estimate of their external diameter being ~46 nm, and they are believed to grow from a nucleation ring of tubulin dimer molecules by addition of further dimers to either surface (Fig. 49). The assembly process has been studied by a variety of techniques including turbidity [294] and small-angle X-ray scattering measurements [296].

5.2.2.2 *Rheological Properties*

Tubulin-containing solutions have not been extensively studied rheologically but some data is available. McIntire et al. [297], for example, have examined viscoelasticity of solutions of microtubules at various concentrations in solvents such as H_2O, D_2O, and glycerol, and have considered the influence of changing the amounts of microtubule associated proteins present (those which influence nucleation), as different distributions of microtubule length should result.

In the dynamic viscoelastic studies, cure curves (Sect. 3.2.1) were measured, and frequency sweep studies were made when these values became constant. For microtubule concentrations in the range 1 to 10 mg/ml, G′ was always small, being of the order of 1 to 10 dyne/cm^2 in aqueous solvents, but values increased in 25% glycerol and 90% D_2O by factors of 3 and 10. In general, modulus components were found to increase with frequency, and G′ was always greater than G″, with the G″/G′ ratio being in the range 0.7 to 0.9. These data tended to be in accord (although not uniquely so) with theoretical predictions for rigid rods, and it was proposed that tubulin solutions could be regarded as behaving in solution as rigid rods interacting by entanglement. Other forms of interaction, of a weak but more specific character, could not be excluded, however.

Like F-actin, microtubule assemblies do not generate strong gel networks on their own. The roles of these materials *in vivo*, it seems, must depend for their success on their participation in an architecture involving other components.

5.2.3 Haemoglobin-S

5.2.3.1 *Introduction*

Whilst the highly organised polymerisations of actin and tubulin occur normally in nature and have important, and beneficial, effects, the same cannot be said of the ordered aggregation of the haemoglobin variant haemoglobin-S(Hb-S) [298]. Normally haemoglobin in red blood cells is unassociated, and functions in oxygen transport in the body, but the variant, when free of oxygen, forms highly ordered structures within the blood cells, and as a result these last become distorted (elongated) in shape. This is the essence of the disease known as sickle cell anaemia, and apparently, the entire phenomenon results because one amino acid in the haemoglobin β-peptide chain becomes incorrectly transcribed [298] during haemoglobin synthesis (a glutamic acid residue gets replaced by a hydrophobic valine residue).

5.2.3.2 *Structural Aspects*

Whilst in the presence of oxygen no polymerisation of haemoglobin-S occurs, the deoxy-haemoglobin-S of sickle cells forms rods several molecules thick (Fig. 49), and these rods are long enough to extend through the blood cell [299]. The rod diameter is of the order of 14.0 to 17.0 nm, and much structural work using X-ray diffraction [298,299], electron microscopy [300], and image reconstruction [301], has been carried out to produce a detailed structural picture.

5.2.3.3 *Rheological Properties*

Little data is available, but Miller, as part of his studies of rod-like particles in solution (Sect. 5.5.1), has examined haemoglobin-S and its polymerisation [302]. He has measured the dynamic modulus of solutions containing haemoglobin-S polymers, and has studied effects produced by changing pH, electrolyte content, and concentration. Experiments were performed at 37 °C and in the absence of oxygen, and it was found that G′ was independent of ionic strength up to a value of 2, and that very little concentration dependence of the modulus occurred in the range 15 gm to 30 gm dl^{-1}.

Miller has suggested that the process of filament formation, and subsequent interaction of filaments to form a network, occurs by spinodal decomposition [303,150], leading to a gel phase of condensed rod-like particles (see later Sect. 5.5.1 on rod-like polypeptide gels). This is in contrast to the behaviour of actin and tubulin in aqueous solution, but it should be recalled that these were studied at much lower concentrations than was haemoglobin-S by Miller. It is not clear how solutions of the Hb-S polymer would behave in a corresponding situation.

5.2.4 Insulin

5.2.4.1 *Introduction*

Insulin is a pancreatic hormone involved in the transport of glucose across cell membranes. The basic subunit has a molecular weight of 5,700 but insulin often occurs as a dimer, or even as a higher aggregate. It is a globular protein with a well-defined native secondary structure, and can be co-operatively denatured in solution, i.e. it has an unfolding temperature.

In its normal function insulin is not involved in large-scale aggregation processes. However, it has long been known that in acid conditions, insulin solutions can be induced to gel and/or precipitate [304,305]. This globular-to-fibrous transition is usually induced by heating insulin solutions above the unfolding temperature (see also similar effects described in Section 5.3.1 for gels from denatured globular proteins). Thus, an opalescent gel can be made by heating a 1% w/w insulin solution at pH 2.0 to boiling point.

5.2.4.2 *Structural Aspects*

Electron microscope [306] studies of insulin gels have revealed a dense fibrous structure (see Fig. 51). Depending on pH, ionic strength, concentration etc., the gel is seen to consist of long protein filaments of varying thickness, entangled and laterally associated, and persisting to great lengths. The thinnest strands are approximately 5.0 nm thick, but multiple strands of much greater thickness can be found, and often one strands winds round another, or diverges from a main strand bundle.

Information about the microstructural make-up insulin filaments, at the molecular level, has come from infrared [305–307], Raman [307], and circular dichroism measurements [305], and from X-ray diffraction [304,305]. The spectroscopic work reveals a substantial increase in β-sheet peptide secondary structure during the gelling process (Fig. 52), and from this evidence, and from fibre X-ray diffraction, Burke and Roug-

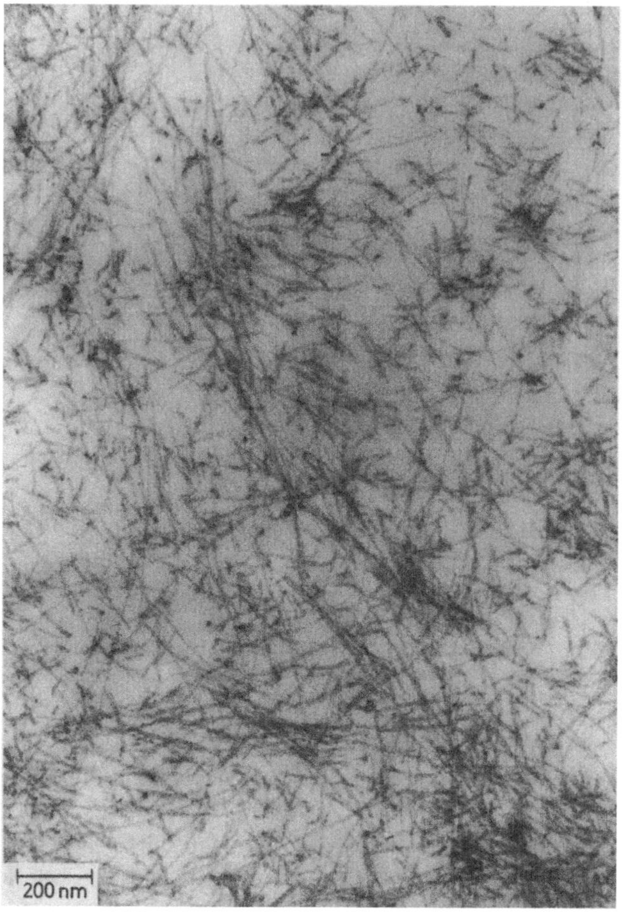

Fig. 51. Transmission electron micrograph of an insulin (heat-set at acid pH) gel section showing rod-like aggregates. Sample was fixed, resin embedded, and stained prior to image recording

vie [305] have proposed a parallel cross-β-structure for the insulin polymer. In this, flat unfolded insulin monomers stack one on top of the other with a chain of hydrogen bonds running at right angles to the molecular plane, i.e. along the fibre axis. In this model, the filaments have hydrophobic exteriors, and this explains their strong propensity to associate laterally as charge-based repulsions are lowered by pH changes, or by changes in ionic strength.

Finally, it is interesting that hormone polymerisation is not confined to insulin. The peptide hormone glucagon also shows this effect at acid pH. The polymerisation is apparently spontaneous, and co-operative, and a "stiff" gel can result [308,309] Tubular filaments appear to form and are capable of lateral association. Again β-sheet is generated as part of the aggregation process, but unlike in the insulin case, no detailed molecular model has yet been established.

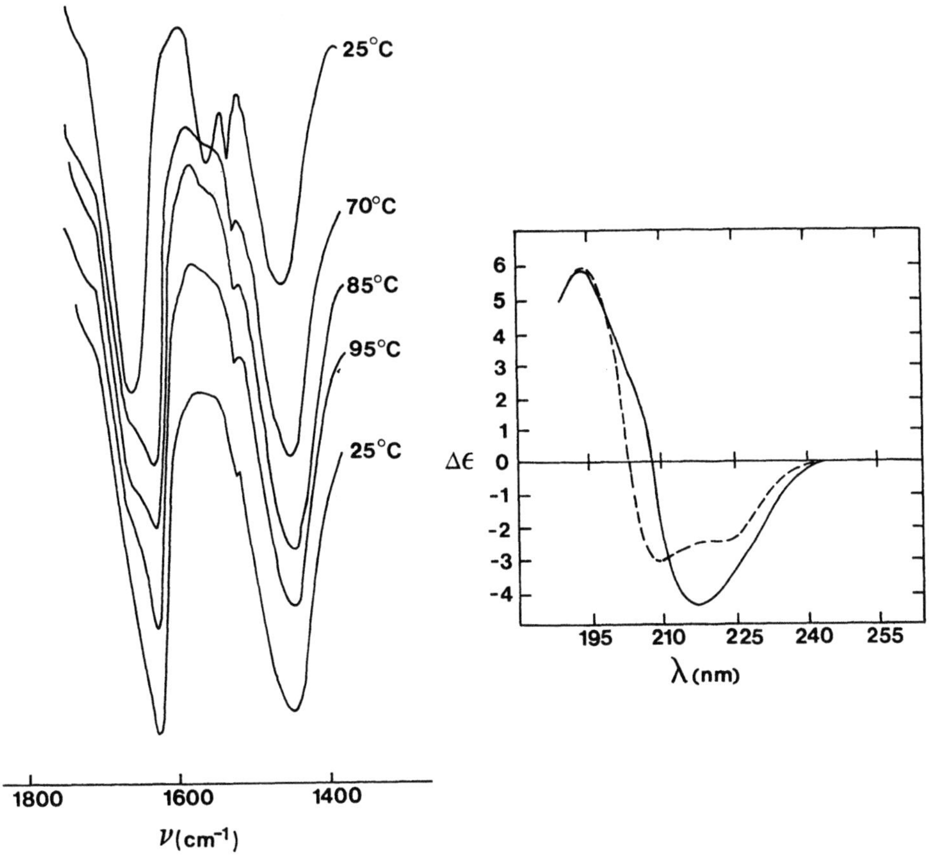

Fig. 52. Infrared and CD evidence showing substantial development of β-sheet secondary structure during insulin aggregation. Reprinted with Permission from Biochemistry *11*, 2435 (1972), Fig. 4, p. 2438. Copyright © 1972 American Chemical Society

5.2.4.3 Rheological Properties

Unfortunately, neither insulin nor glucagon gels have been characterised rheologically. Because of their structural similarity in gel form, to the fibrin clots to be described next, however, it is possible that they behave rheologically very much like these.

5.2.5 Fibrin

5.2.5.1 Introduction

Another important example of the organised aggregation of a compact protein to form fibrous structures, and ultimately a three-dimensional gel network, is provided by the polymerisation of the blood protein fibrin, which is itself generated by action of an enzyme thrombin on a precursor protein fibrinogen (natural blood coagulation mechanism)[3,310,311].

In the present Section the process of fibrin assembly will be discussed in detail. First, however, some comments are appropriate about the precursor fibrinogen from which fibrin derives.

5.2.5.2 Fibrinogen

Normally, in cases where protein gelation is being considered, the polymerising monomer is well characterised, and the nature of the aggregation process is the object of research. This is not true of the precursor of fibrin gels, fibrinogen. Earlier reviews [310, 311] have discussed uncertainties about the size and shape of this protein, and as a fairly recent article by Müller and Burchard [312] makes clear, in this respect the position has not changed much in the ensuing decade. There is still controversy about fibrinogen, many authors having attempted to characterise fibrinogen structurally, using many different techniques.

Most of the models proposed have been rod-like, but conclusions about the rod length, and thickness, have varied considerably. In addition, globular models have been proposed, such as the spherical structures of Koppel [313] and Marguerie [314]. Techniques used to arrive at these structures have included sedimentation, flow birefringence, viscoelastic measurement, viscosity, light scattering, electron microscopy, neutron scattering, dynamic light scattering, and X-ray scattering.

Despite the sophistication of some of the techniques just listed, the most celebrated and widely accepted model for fibrinogen comes simply from electron microscopy, and is that proposed in 1959 by Hall and Slayter [315]. These authors concluded that fibrinogen has a nodular, extended structure (Fig. 53) consisting of three globular particles assembled in a row (two 6 nm spheres enclosing a smaller 5 nm sphere) and connected by peptide threads. This model has also led to a popular description of fibrin protofibril formation (the initial step in fibrin aggregation) in terms of the assembly of a staggered double run of trinodular species (Fig. 53). Such a model is said to explain the X-ray diffraction lines observed for fibrin gels by Stryer [316] (and subsequently by other authors) [317] and a well documented striation pattern observed for fibrin gels by several electron microscopists [313, 318, 319].

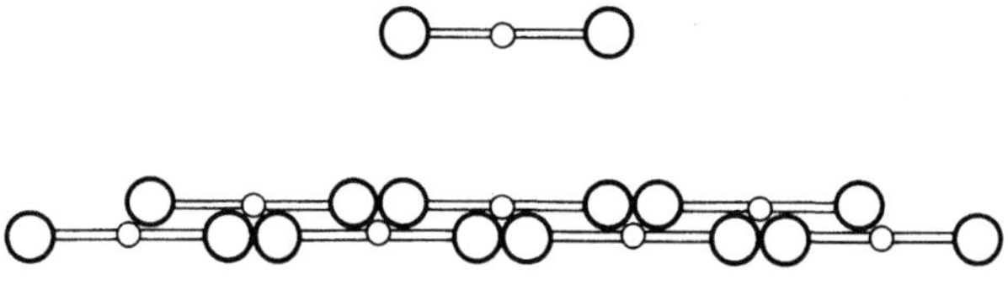

Fig. 53. Representation of Hall & Slayter model (Ref. [315]) for fibrinogen and fibrin

The trinodular model for fibrinogen of Hall and Slayter has dimensions of approximately 45 nm by 9 nm, and the half-staggered assembly process in fibrin leads to a predicted X-ray repeat spacing of ~22.5 nm (half this length) which is roughly what is observed experimentally. As the article by Müller and Burchard [312] makes clear, however, the matter is still open to doubt, for in a list prepared by these authors of all of the rod models proposed for fibrinogen, rod structures are found ranging in length from 90 nm to 45 nm, i.e. differing by up to a factor of two. Müller and Burchard point out that all experiments done in aqueous solution lead to rod-lengths greater than 45 nm, and they note that X-ray scattering studies by Lederer and Hammel [320] have led to a rod length of 54.4 nm and to a molecular weight of 335,000. This is consistent with a mass per unit length (M/L) of 6150 nm^{-1}, and it is interesting that in Lederer and Hammel's work an occasional M/L value as low as 4200 nm^{-1} was obtained, and a corresponding rod length of 80 nm.

In an effort to resolve the uncertainties regarding fibrinogen structure, Müller and Burchard [321] have performed accurate static light scattering measurements for human and bovine fibrinogen in solution under physiological conditions, and have found a molecular weight of 340,000, and complex conformational behaviour [321,322]. Holtzer plots (Fig. 54) of the data, for example, could only be fitted using models assuming a mixture [312,322] of short (Hall-Slayter) and long rod components. In detail, the fits achieved suggested a considerable variation between the bovine and human samples, as the bovine data implied a mixture of 21% long rod component, whilst the human data could be fitted to 71% of the long rods. Evidence for this conclusion was derived from dynamic light scattering experiments [322] (see also Sect. 2.3.2), and it was concluded that fibrinogen in solution could exist as a mixture of isomeric forms (Fig. 55). This was offered as an explanation [312,321] for the variety of structures proposed for fibrinogen over the years, including the so-called banana model of Marguerie [314], the short folded rod of Bachman [323], and Müller and Burchard's

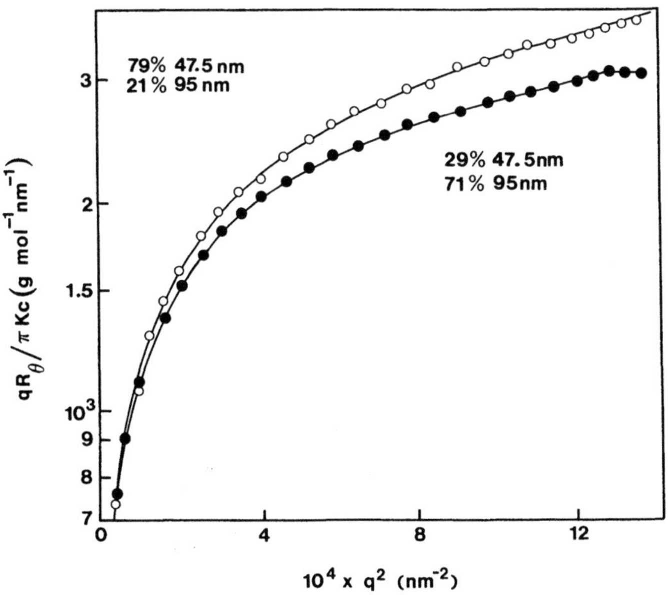

Fig. 54. Holtzer plots of integrated light scattered intensities from human (●) and bovine (○) fibrinogen measured under identical conditions of pH, concentration, ionic strength and temperature. Solid lines are theoretical fits based on mixtures of rods of two lengths (47.5 and 95 nm) but identical molecular weights. Mass fractions giving best fits are indicated. Reproduced fröm Müller, M. and Burchard, W. International Journal of Biological Macromolecules, 1981, 3, 71–76, by permission of the publishers, Butterworths & Co. (Publishers) Ltd. ©

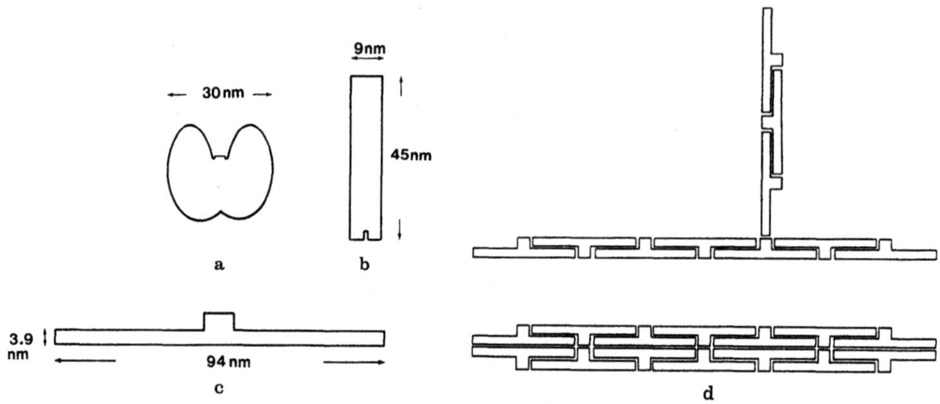

Fig. 55a–d. Representations of three suggested isomeric forms (**a–c**) for the fibrinogen monomer (Ref. [321]). Also shown (**d**) is a scheme for fibrin polymerisation based on the most extended rod form (**c**). This model has been proposed (Ref. [321]) as a result of light scattering experiments

own extended rod, which was described as having a length of 94 nm, and a width of 3.9 nm. These two authors also concluded, (a) that fibrin was not dramatically different in structure from fibrinogen, and (b) that it is the elongated fibrin rod of length 90 nm which is incorporated into fibrin protofibrils. It should be added, however, that, whilst the light scattering technique is likely to be one of the most suitable approaches for studying fibrinogen, and fibrin aggregation, the trinodular (short rod) structure of Hall and Slayter still finds substantial support in the literature, this model having, for example, being adopted quite recently by Roska et al. [317], and is supported by microscopy studies [322].

5.2.5.3 Structural Studies of Fibrin and Fibrin Polymerisation

As in the case of fibrinogen, structural approaches such as electron microscopy, X-ray diffraction/scattering and light scattering, have all been applied in the study of fibrin, and the networks formed from it. It seems to be generally agreed that the transformation by enzyme action of the fibrinogen monomer to the fibrin monomer, is a small one. Two peptide chains (A and B) are removed from fibrinogen to expose reactive sites, whence aggregation of fibrin monomers to form fibrin polymers, proceeds steadily unless inhibited, and eventually gelation occurs.

Early X-ray diffraction studies of such gels by Bailey [324], using a wide-angle approach, led to the conclusion that fibrin molecules, as monomers, were structurally similar to fibrinogen, and further experiments using light scattering, and X-ray scattering/diffraction, have not altered this conclusion. Müller and Burchard [312] have made very detailed studies of fibrin polymerisation by light scattering, and have concluded using Zimm, Holtzer (Fig. 56), and Kratky plots that the aggregation of fibrin monomers initially involves the formation of elongated rod-like species based on their extended structure for fibrinogen not the shorter Hall-Slayter structure. After this initial linear aggregation, a degree of branching occurs and, depending on conditions, a degree of lateral association (cf. Fig. 55).

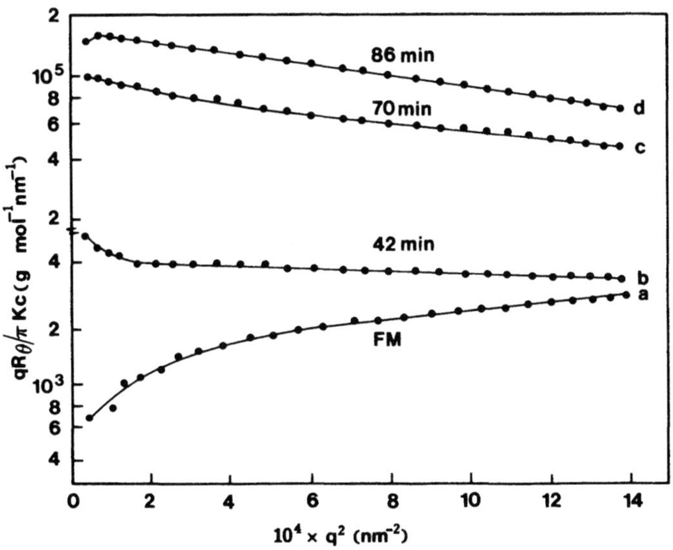

Fig. 56. Holtzer plots for the human fibrinogen monomer and for fibrin aggregates occurring at various times after initiation of the polymerisation process. Initially rod-like structures form (curve b) but later (curves c and d) there is evidence of branching. Reproduced with Permission from Biochim. Biophys. Acta *537*, 208 (1978), Fig. 5, p. 217

In a very elegant application [325] of cascade theory, these workers have shown that their light scattering data for fibrin polymerisation can be explained quantitatively by assuming the phases of aggregation just mentioned. Changes in the conditions of gelation, and the source of the fibrin, were found to influence the exact details of branching and lateral association which occurred, but essentially, the same model could be applied in all cases.

Apart from light scattering, other physical approaches have been used to study fibrin gelation, and have provided structural details. Electron microscopy, for example, has provided dramatic images [311,326,327] showing fibrous structures, the fibre thicknesses showing great variability, but a constant characteristic striation pattern as discussed above. Fine transparent clots, not unexpectedly, showed fibres of low cross-sectional width, whilst coarse opaque clots were less uniform, and had much higher average fibre diameters. The network structures indicated by the microscope approach are closely similar to those which have been obtained for the heat-set insulin gels discussed in the last Section, fibrils tending to entwine, as well as showing evidence for lateral aggregation and branching. Some of these features are particularly clearly demonstrated in images obtained recently during a very comprehensive transmission electron microscopic study [327] of fibrin gels and films (Fig. 57).

5.2.5.4 *Rheological Studies of Fibrin Gels*

In 1947 Ferry and Morrison [328] published an early report on the mechanical implications of the fibrinogen-fibrin conversion. Mechanical measurements were of several kinds, and included (a) estimates of gel rigidities using a wave propagation method;

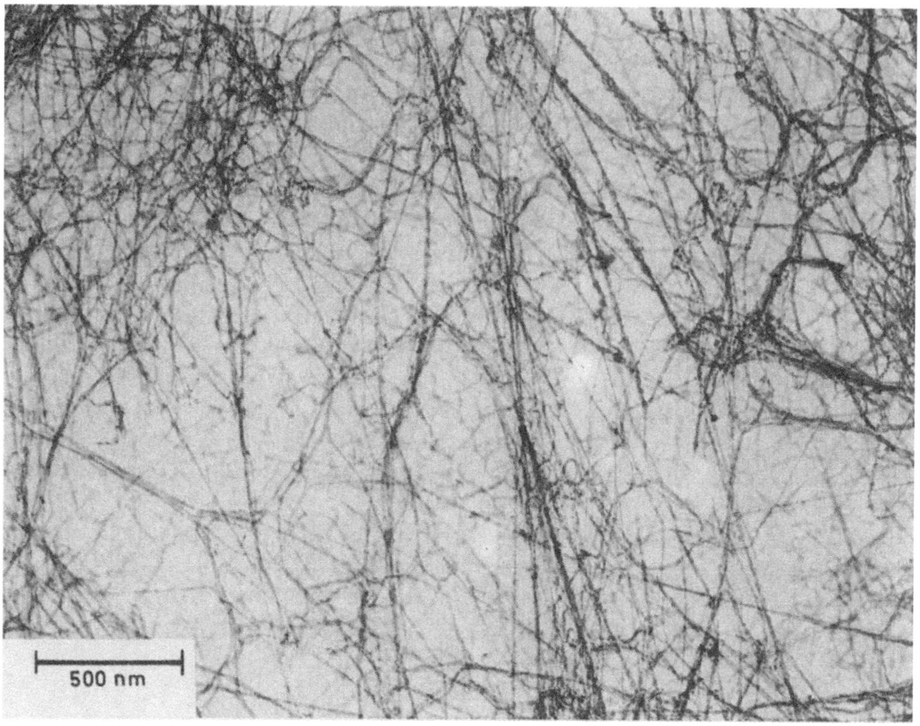

Fig. 57. Transmission electron microscope image of a fibrin network formed under conditions of limited lateral aggregation of fibrin chains. There is evidence of cross-linking via branching and chain entwining mechanisms. Reproduced with Permission from J. Mol. Biol. *174*, 369, (1984), Fig. 2, p. 374

(b) estimates of so-called gel friability and syneresis, by qualitative observation of gel response under pressure; and (c) measures of extensional and fracture properties of fibrin clots under applied load. Experiments performed using these methods were directed towards determining the consequences for gel rheology of changing gelling conditions such as the polymerisation temperature, the pH, the ionic strength of the polymer segments, and the concentration of fibrinogen and thrombin originally present in solution. A number of conclusions were reached including the following.

First, it was found that after thrombin was added to a fibrinogen solution opacity started to increase, and, after an appropriate "clotting time", the solution became rigid. The rigidity and opacity continued to increase through the gel point, but the rates of increase fell off, and eventually limiting values were approached. Fibrin gels, formed in this way, were classified as "fine" or "coarse" depending on their physical properties. Fine clots, for example, made under conditions (pH, ionic strength) leading to high protein-protein repulsion were transparent, elastic, and friable and were also non-syneresing with a low elongation at break. Coarse clots, on the other hand, were opaque, plastic, and non-friable, and syneresed readily.

Moduli of rigidity, measured for fine clots made at different fibrinogen concentrations, varied (approximately) with the 1.6 power of the concentration, and values were higher than could readily be accounted for by assuming a rubber-like front factor (Sect. 3.4).

In 1951, Ferry, Miller and Shulman [329] published further observations on the mechanical properties of fibrin gels. In this case, attention was focused on gels from bovine fibrinogen, not human fibrinogen as in the previous study, and measurements of gel rigidity were again made using the wave propagation approach. This high frequency, short-time measurement was supplemented by stress-relaxation observations. The rigidity of fibrin gels was studied as a function of concentration, and at various temperatures after the network had become established, and a comparison of gel properties made in both the presence and absence of calcium ions.

The studies of the temperature dependence of rigidity showed clearly that gel rigidity decreased with increasing temperature, and also showed that the change in modulus with temperature was reversible. The concentration studies revealed a somewhat different power law in relation to concentration than had been found previously for human fibrinogen (a 2.3 rather than 1.6 power law). It was noticed that the bovine fibrin gel exponent varied with temperature, and this property was contrasted with the apparent temperature invariance of the corresponding gelatin exponent (~ 2.0).

In terms of fibrin gelation in the presence of calcium ions, a highly significant effect was observed. The calcium gels showed much slower stress relaxation indicating the presence of more permanent bonds in the network, and at comparable fibrin concentrations, the calcium gels were generally the more rigid. The moduli of calcium gels were also found to be less temperature sensitive. Calcium-free gels showed rapid stress relaxation and were weaker. It was concluded from these observations that calcium ions in the presence of an appropriate serum factor, called fibrinoligase (present in the fibrin preparations used), introduced more permanent cross-links such as covalent bonds, and hence altered the gel rheology substantially. This conclusion was in line with other observations on the effect of calcium ions on blood coagulation, since the importance of calcium in relation to the clotting mechanism had long been recognised [311].

Over the years Ferry and co-workers have continued to research the mechanical properties of fibrin gels and a number of contributions by them have appeared in the comparatively recent literature. In 1974, for example, as part of this effort, Roberts et al. [330] described measurements of the storage and loss components (G', G'') of the viscoelastic shear modulus for fibrin clots from purified human fibrinogen. A modified Birnbaum transducer apparatus [11] was used, and frequency dependence established for the modulus in the range 0.01 to 160 Hz. Clots were prepared under various conditions of pH and ionic strength and, as in previous work, these were described as "fine" or "coarse" depending on the level of network heterogeneity and resultant opacity. Fine unligated clots (not covalently crosslinked with fibrinoligase) showed little mechanical loss, or G' frequency dependence, but it was concluded that loss mechanisms probably contributed at frequencies above the highest frequency accessible in the experiment. For these fine clots, G' was found to be proportional to $C^{1.5}$. Coarse unligated clots, also studied in this work, showed a small increase in G' with frequency indicating relaxation mechanisms with reciprocal time constants within the measurement frequency range.

In further studies of the unligated gels, Ferry and co-workers, extended [331] the time scale of measurement using creep experiments. By combining the new data with that described in the last paragraph, a description of viscoelastic behaviour was achieved over seven logarithmic decades of time (or equivalently, frequency). Fine, unligated clots, did not creep significantly over this time scale (longest time 10^4 secs.), but the coarse unligated clots did. It was concluded that whilst the fundamental protofibrillar elements of the networks remained intact under stress, the bundles of fibres present in coarse gels showed a tendency to slip. Interestingly, it was found also that ligation suppressed this effect, presumably by securing the bundles of protofibrils. Additionally, studies of the strain dependence of the creep experiment results, and tests of creep recovery by the Boltzmann superposition criterion, suggested conformity with linear viscoelastic behaviour, and indicated that structural changes had not been induced in the gel networks during measurement. Later work in 1976, also employing creep measurements led to similar conclusions [332)], and the suggestion was made that the fine, unligated, clots, based on largely unassociated protofibrils, owed their elastic behaviour to a steric blocking of these long stiff protein polymers.

The origin of elasticity in fine unligated clots is an intriguing topic and it has been further investigated and discussed by Ferry and co-workers. Nelb et al. [333)], for example, in 1980, pursued this problem by performing more modulus, creep and creep recovery experiments. In this way, fine unligated clots from both human and bovine fibrinogen were examined, and an exponent of 1.90 was found for the modulus-

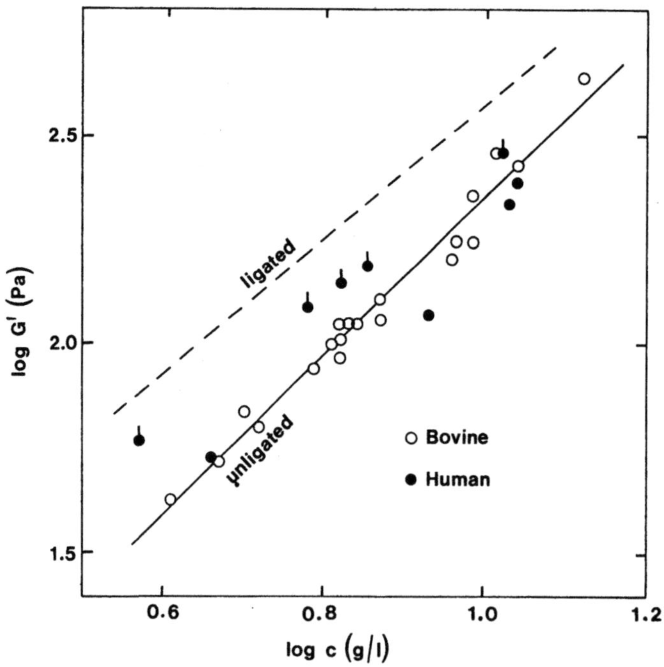

Fig. 58. Double logarithmic plot of modulus-versus-concentration data for ligated and unligated fibrin clots from human and bovine fibrinogen. The exponent n in the relationship $G' \propto C^n$ may be obtained for each system as a straight line slope. Reproduced with Permission from Biophysical Chemistry. *13*, 15 (1981), Fig. 1, p. 17

concentration dependence (Fig. 58). Nelb et al. concluded that this power law was inconsistent with the earlier idea that the viscoelastic behaviour of the fine unligated clots owed its existence purely to steric effects, although the theoretical treatment of the steric interaction of rods [105] suggested that a much higher power law would apply. Nelb et al. did not provide a full explanation for the effects of bending on fibrin elasticity, however, and it was left to Janmey et al. [334] in 1983 to provide what is probably the most realistic hypothesis currently available. They accepted that lateral junctions exist between protofibrillar elements, even in fine clots, and hence that the interactions between these are not merely steric. They explained elasticity largely in terms of energy storage as protofibrils bend, though they also accepted that a steric contribution of the Doi and Kuzuu form might be present. Their description of the connectivity of the gel network in fine unligated clots has found graphic confirmation in recent electron micrographs [327]. Finally, it is worth noting that exponents of less than two in the modulus — concentration relation would occur quite naturally for systems following a low-functional branching process. This point is examined in more detail in Section 3.4.

5.2.5.5 Conclusions

The fibrinogen-fibrin system provides an excellent example of a biopolymer gel network, and the history of its investigation is illuminating. The phenomenon involved is quite unlike the formation of cold set gels from gelatin and polysaccharides (Sect. 4) and is much more akin to the heat-set gelation of insulin described in the last Section. There are also elements of similarity to the formation of fibrillar structures in actin and tubulin solutions, but as the previous descriptions of these systems make clear, in fibrin, the chain-chain interactions are much stronger, and in consequence, more rigid and permanent networks are formed.

5.3 Globular Proteins Forming Branched Networks

5.3.1 Gels From Thermally-Denatured Globular Proteins

5.3.1.1 Introduction

This Section considers the form of protein association which occurs when denatured globular proteins interact to form aggregates and gel networks. In some situations, relatively homogeneous networks of twisted, branched, filaments arise. In others, the gel structures produced show great heterogeneity. In all cases, however, there is little evidence for the long linear protofibrillar structures found in insulin and fibrin gels or in the temporary actin and tubulin networks.

Generation of gel networks from solutions of globular proteins usually requires a degree of unfolding of the protein as a first step, and this can be achieved thermally or chemically. Heat-set gels formed by thermal denaturation and aggregation of globular proteins in solution, are very easily made, and indeed most globular proteins can be induced to behave in this way if solvent conditions, such as pH and ionic strength, are appropriate. The process is of practical significance in food preparation, both in the home and in industry, as network formation by globular proteins can dramatically contribute to food texture (as, for example, in the preparation of a boiled egg) and to water binding, emulsion stabilisation, and flavour retention [335].

In view of its practical importance and convenience, the remainder of the present Section will deal with gelation by heat-setting. Chemically-induced unfolding and gelling of globular proteins will be discussed separately.

5.3.1.2 Structural Aspects

Early structural investigations of heat-set globular protein gels made use of X-ray diffraction [336], and examined fibres prepared by a combination of heating and stretching protein solutions. This work led to a description of the denaturation process as involving substantial, if not total, loss of native tertiary and secondary structure; and of the aggregation process as involving a subsequent packing of peptide strands into ordered junction zones based on the β-sheet conformation.

The first real departure from this picture was by Barbu and Joly [337] in 1953. In a study of thermally-induced aggregate formation by horse serum albumin, and certain other proteins, Barbu and Joly concluded that the aggregates produced arose by polymerisation of partially unfolded protein molecules still in corpuscular form. From electron microscope evidence, and from studies by streaming birefringence, they also concluded that two types of aggregate could be made, depending on protein charge, i.e. relatively linear and unbranched polymers when repulsion was substantial, and highly branched clusters of protein particles when it was small.

Some years later, a similar model was proposed by Kratochvíl and co-workers [338–341] for aggregates formed from human serum albumin (HSA). Light scattering experiments were performed on dilute protein solutions, and molecular weight information was derived as a function of time, as aggregation proceeded. This time-dependent data was analysed using a kinetic model [339] in which unfolded HSA molecules were assumed to be corpuscular, and to have f binding (i.e. cross-linking) sites of comparable strength. An equation was derived to relate weight average molecular weight to time, using Flory's theory of gelation [110], combined with a second-order rate equation relating the rate of disappearance of binding sites to free binding site concentration. The result was,

$$M_w = M_0(1 + 2fC^0 kt)/(1 - f(f - 2) C^0 kt) \tag{34}$$

in which M_0 was the monomer molecular weight, $f > 2$ was as defined above, C^0 was the initial monomer concentration, k a rate constant, and t the time. This model predicts gel formation, since divergence of the molecular weight occurs at a gel time,

$$t_g = 1/f(f - 2) C^0 k \tag{35}$$

By applying this approach to light scattering data, Kratochvíl et al. [339] found a value for f lying between 2 and 3 (for HSA and a particular set of gelling conditions) and the fractional result obtained was assumed to indicate polydispersity of the reaction sites.

Like Barbu and Joly [337], Kratochvíl and co-workers also distinguished between the tendency towards linear aggregate formation when repulsive forces were large, and so-called "random" aggregation when repulsion was low. Whilst Barbu and Joly described the protein aggregation as reversible, however, Kratochvíl et al. saw it as essentially irreversible.

In more recent years, continued application [342] of experimental techniques such as electron microscopy, viscometry, and osmometry has tended to support the corpuscular model, as has use of other approaches such as small-angle X-ray scattering [35,343]. This last was applied fairly recently by Clark and Tuffnell to study the aggregation and gelation of bovine serum albumin (BSA) under a variety of conditions of pH, ionic strength and protein concentration. Using the angular dependence of X-ray scattering in the Bragg spacing range ~40.0 nm to 1.5 nm, for both BSA solutions and gels, these workers confirmed the corpuscular character of the aggregation process suggested by early work and found a characteristic scattering maximum for gels made at high pH and low ionic strength, i.e. under conditions of strong inter-protein repulsion. This maximum was absent in scattering from the more heterogeneous networks which arose under conditions of reduced repulsion (lower pH, high salt content). The X-ray scattering results for BSA aggregates and gels were modelled by computer simulation, and it was concluded that the gels showing scattering maxima were based on chains of BSA molecules forming a protein network of extremely uniform pore size distribution. The thickness of the chains was estimated to lie between one and two native BSA monomer diameters (~ 10 nm) but the detailed arrangement of the monomers in the networks strands was not established. The change in X-ray scattering as the gels became opaque was assumed to indicate a transition to a network with a vastly more polydisperse distribution of pore size, and more frequently branched network strands.

Further structural information about heat-set globular protein gels has come from electron microscopy which has been applied in various forms including transmission studies, scanning electron microscopy, and freeze fracture. Hermansson and Buchheim [344], for example, have applied all of these methods in a systematic study of gels based on the soya protein glycinin, and have compared and contrasted the various approaches. These authors have concluded that, provided great care is taken to avoid artefacts (particularly during freezing), consistent results can be obtained.

Clark and co-workers [306] have also explored the electron microscope approach, and have examined a range of protein systems including the BSA gels studied by X-ray scattering. They confirmed that BSA gels made under conditions of high protein charge did indeed have highly uniform network structures, (Fig. 59) and that opaque gels did not, and they demonstrated that network strands in both types of gel were of the order of one to two molecular diameters in thickness, and had an apparently monodisperse size distribution. This meant that in cases where the network structure was heterogeneous, there was little evidence for substantial lateral association of network strands, the average frequency of branching being simply increased. In this work gels from several globular protein types were examined, and in cases where a comparison with results from small-angle X-ray scattering was possible, no serious contradictions emerged.

The change from uniform network structure to inhomogeneity, which often occurs quite sharply for a gelling globular protein as gelling conditions are changed, is of great interest, as it is usually accompanied by substantial alterations in gel properties. In such situations, for example, protein gels become turbid and water-releasing, and though they may also become intrinsically more rigid, they are also likely to become more brittle. The changes in underlying network uniformity giving rise to these effects are associated with changes in the balance between attraction and repulsion effective for the aggregating particles, for they are usually generated by changes in pH and/or ionic strength.

For globular protein gels formed by heating, thermodynamic equilibrium is unlikely but, experimentally, metastable phase diagrams can be constructed showing a boundary between optically clear solutions and gels, and turbid gels and suspensions of microgel particles. A typical result [35] appears in Fig. 60, a line separating gelled from ungelled solutions also being shown. This indicates the establishment of a miscibility gap region, and for BSA gels, this gap can change radically in position as pH and/or ionic strength are altered. Turbid solutions and gels arise within the gap, opaque solutions (or suspensions) becoming continuous gels as the protein concentration is increased. It is interesting that the situation observed is consistent with that predicted by the Coniglio, Stanley and Klein [139] model discussed earlier (Sect. 3.6).

5.3.1.3 Cross-linking Mechanisms and Secondary Structure Change

In general, the interactions between unfolded globular proteins are likely to be complex, as a whole series of attractive mechanisms can operate, such as hydrogen bonding, ionic interactions, hydrophobic forces and even covalent bonding. The disruption of the stable native form, on heating a globular protein in solution, makes all of these forces potentially operative, and one reason why exact details of cross-linking have rarely been established for this type of network is the diversity of interactions which may be present simultaneously. As Clark and Lee-Tuffnell have pointed out in a recent review [35] of globular protein gelation, most experimental approaches to elucidating the cross-linking mechanisms are fraught with ambiguities, and a clear-cut separation of factors is rarely achieved.

One molecular aspect of the aggregation process which can be pursued fairly easily, however, is the question of peptide secondary structure change during gelation, and its relation to cross-linking. Early descriptions [336] of thermally-induced globular protein network formation, it will be recalled, attached great importance to specific secondary structures as the origins of chain association, and although this early, and simplistic view, has largely been abandoned, it may still, in a limited sense, contain elements of truth.

The experimental approaches available for measuring secondary structural content (α-helix, β-sheet and random coil components) of proteins have already been discussed (Sect. 2.2). They are principally the methods of infrared, Raman, and circular dichroism spectroscopy. Experiments using these approaches usually seek to quantify secondary structure in both the unheated solution and in aggregates or gels formed by heating, and Clark and Lee-Tuffnell have discussed [35] a whole series of such

▶

Fig. 59 a–c. Transmission electron micrographs (two magnifications) for bovine serum albumin (BSA) gels showing varying degrees of network heterogeneity. **a** images for a pH 6.5, 10% w/w clear gel in distilled water. **b** a pH 6.5, 10% w/w turbid gel in 125 mM NaCl, and **c** a pH 5.1 10% w/w coagulate at the isoelectric point. Reproduced with Permission from Int. J. Peptide Protein Res. *17*, 380 (1981), Fig. 1, p. 384, Copyright © 1981, Munksgaard International Publishers, Copenhagen, Denmark

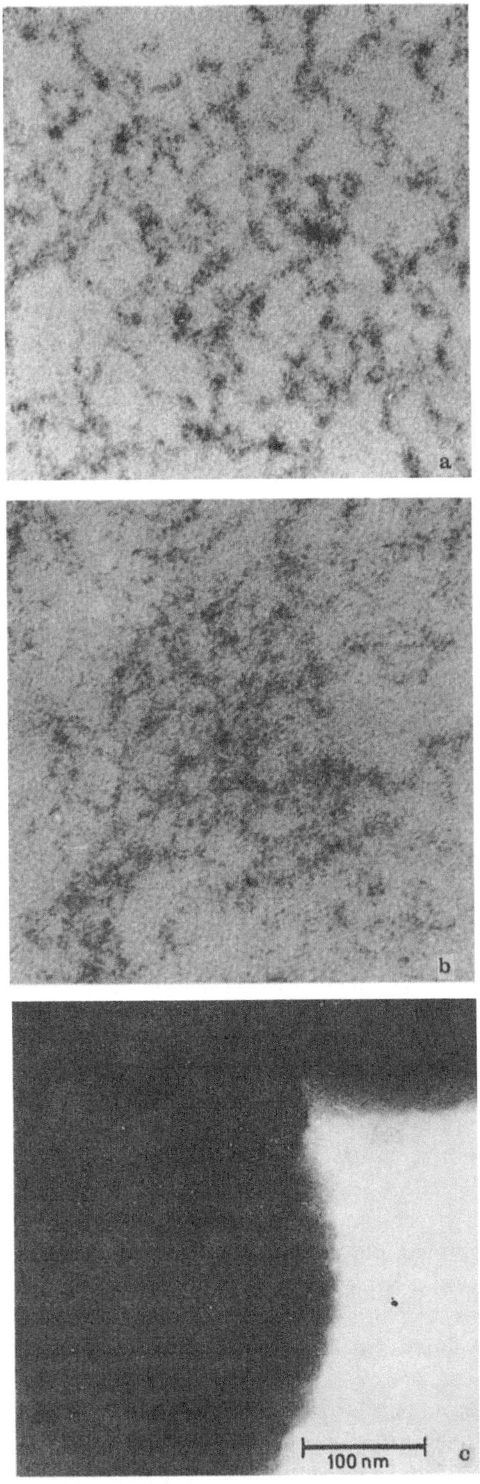
a
b
100 nm
c

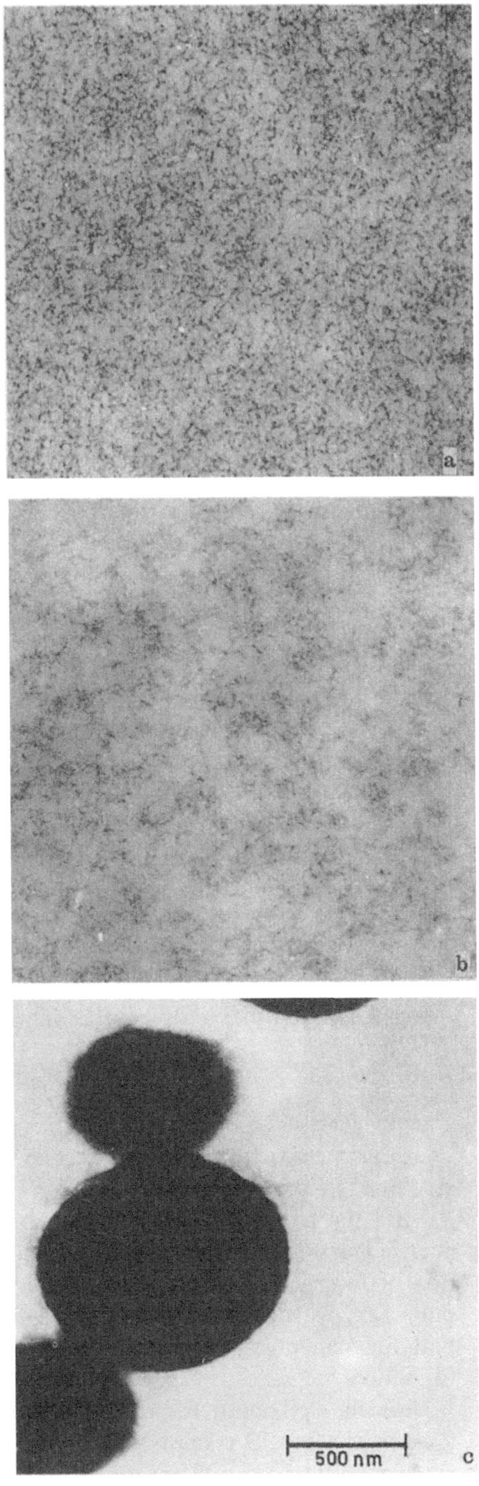
a
b
500 nm
c

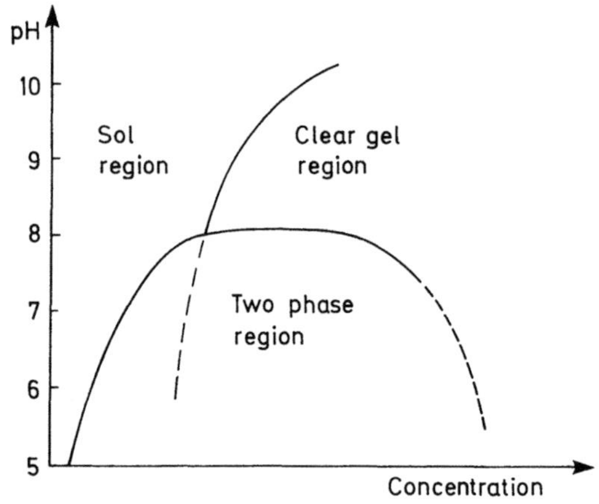

Fig. 60. Qualitative indication of phase behaviour of BSA gels as pH and protein concentration are varied at fixed ionic strength. The broken line represents extension of the sol-gel line into the two-phase area. Reproduced with Permission from "Functional Properties of Food Macromolecules", edited by J. R. Mitchell and D. A. Ledward, Elsevier Applied Science, 1986, Fig. 30, p. 265

experiments applied to various protein systems. The series included the BSA gels studied by other techniques, and also gels from ribonuclease, lysozyme, α-chymotrypsin and the egg protein, ovalbumin.

In summary it may be said that the gels invariably have different secondary structures from the native proteins but it is found that the difference varies greatly from system to system. Bovine serum albumin gels, for example, have somewhat less α-helix in the gel form than in the original sol, but much more β-sheet and disordered conformation. In the ovalbumin case similar effects are observed but the generation of β-sheet on aggregation is much more pronounced.

For the BSA and ovalbumin system, detailed studies of the sensitivity of these secondary structure changes to gelling conditions and to protein concentration, have indicated little functional dependence, since it is found [35] that the proportion of β-sheet in the gel state is roughly the same, whatever the degree of protein cross-linking. This indicates that, if the sheet structure plays a role in cross-linking, this role is very limited, involving the formation of protein dimers, for example, which subsequently generate the network. On the other hand, the sheet may arise intramolecularly, and may only have an indirect influence on protein-protein bonding.

5.3.1.4 Rheological Properties

Rheological studies of heat-set globular gelation divide naturally into two categories, linear viscoelastic studies, and studies at higher applied deformations, including studies of failure behaviour.

Until the late Seventies, detailed linear viscoelastic studies of heat-set globular protein gelation seem not to have been attempted, but in 1981 an extensive investigation of the rheological properties of heat-set BSA gels at small deformation was published by Richardson and Ross-Murphy [61,83]. This work was concerned with both fully-cured gels at concentrations well above critical, and with incipient network formation.

Near the gel point, for example, measurements [61] of gel times (falling sphere viscometer) for BSA solutions at pH 6.5, as a function of temperature, revealed protein unfolding as rate limiting below 57 °C, and aggregation as rate limiting above this temperature. In addition, the effects of protein concentration on the gel time, and on the growth profile of G′ as a function of time, were studied for a series of BSA solutions at 59.4 °C. A torsion pendulum was used in this work, and by extrapolation to "infinite gel time", a critical concentration for gelation was established for the BSA pH 6.5 system of ~7% w/w. By combining the modulus and gel time data with optical rotation changes assumed to be related to cross-linking, and combining the measured gel times with values of the optical rotation at the gel point and at infinite time, the critical degree of cross-linking at the gel point was established. This enabled weight average values for the BSA functionality f_w to be calculated, and results lay in the range $2.5 < f_w < 2.05$, in agreement with Kratochvíl's earlier findings for HSA [339].

Finally, in this study, Richardson and Ross-Murphy compared experimental values for G′ measured beyond the gel point with theoretical predictions based on cascade theory [112] estimates of the density of elastically active network chains, and concluded that the contributions these chains made to the gel modulus were much greater than could be explained solely by invoking the classical theory of rubber elasticity [78]. Strong enthalpic contributions to the moduli of BSA gels were suggested.

In a second paper [83] on the BSA system, Richardson and Ross-Murphy examined fully-cured gels made under varying conditions of pH, sodium chloride content, and protein concentration. In this investigation, both G′ and G″ (1 Hz) were measured as functions of time (torsion pendulum) during cross-linking at 95 °C, and then as functions of frequency (0.001 to 16 Hz; Rheometrics Mechanical Spectrometer) temperature and strain, on completion of curing. The essential conclusions about BSA gelation were as follows: (a) for a given set of gelling conditions, the gel formed

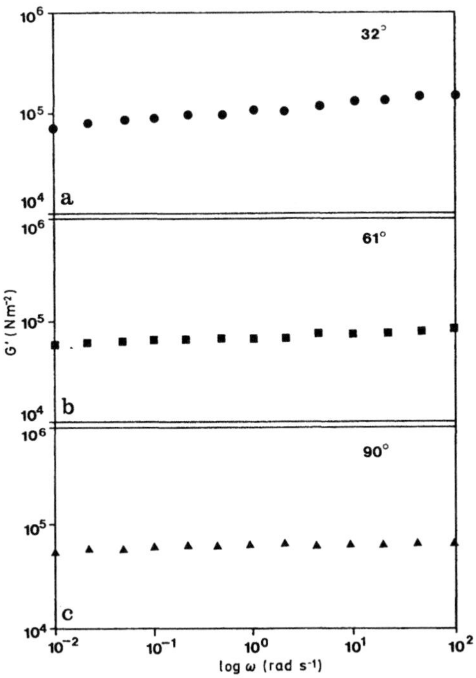

Fig. 61. G′ — frequency data for BSA gels (15% BSA, pH 6.0, 10 mM added NaCl) at three temperatures. Reproduced with Permission from British Polymer Journal, *13*, 11 (1981), Fig. 4, p. 13

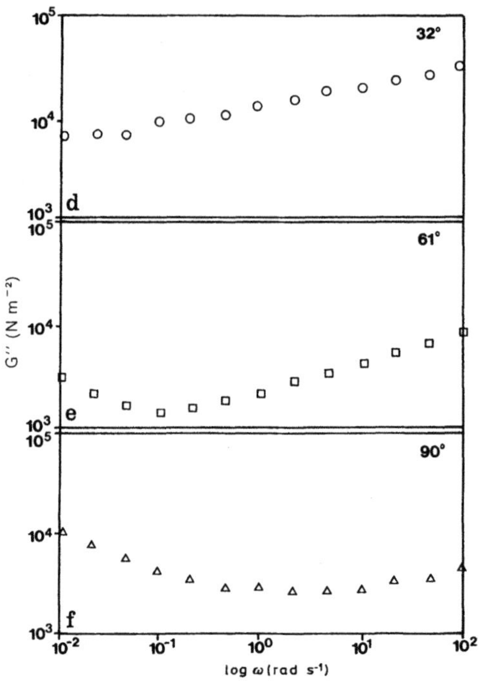

Fig. 62. G'' — frequency data for the systems and conditions of Fig. 61. Reproduced with Permission from British Polymer Journal, *13*, 11 (1981), Fig. 4, p. 13

behaved as a rigid elastic solid, provided that protein concentration exceeded a critical value C_0: (b) modulus-versus-frequency results indicated that log G' varied linearly with log ω, and showed only a small increase with frequency (see Fig. 61): (c) log G'', on the other hand showed a slight minimum (Fig. 62) at a frequency which changed with temperature: (d) both G' and G'' fell with increasing temperature and rose to their original values on cooling: (e) values of G' (1 Hz) measured for a whole range of gelling conditions, showed that as the gelling pH increased (6.0 to 8.0) the effect of ionic strength on the modulus became crucial. At the highest pH, the critical concentration varied greatly with sodium chloride content, whereas at the lowest pH, for the turbid gels formed, the influence of salt was much less: (f) measurements of G''/G' ratios for the different gels showed that whilst G' was generally much greater than G'', the opaque gels, though more rigid, were also less "elastic" than the high pH transparent systems: (g) for the gels studied, the rheological properties were relatively insensitive to strain, linear viscoelastic behaviour being found up to, and beyond, strains of 0.1.

In this work on fully-cured BSA gels, Richardson and Ross-Murphy also examined the important issue of the concentration dependence of the gel modulus (see Sect. 3.4). A plot of log G' against concentration for BSA gels cured at 85 °C had a characteristic form (Fig. 63) very similar to corresponding data published earlier for the soya component glycinin by Bikbov and co-workers [345,346]. These latter had drawn attention to the fact that a theoretical function, derived by Hermans [114] for weakly cross-linked polysaccharide systems, could be fitted to their data, and Richardson and Ross-Murphy found the same to be true for BSA. Subsequently the Hermans' approach has been extended to describe the form of the modulus concentration

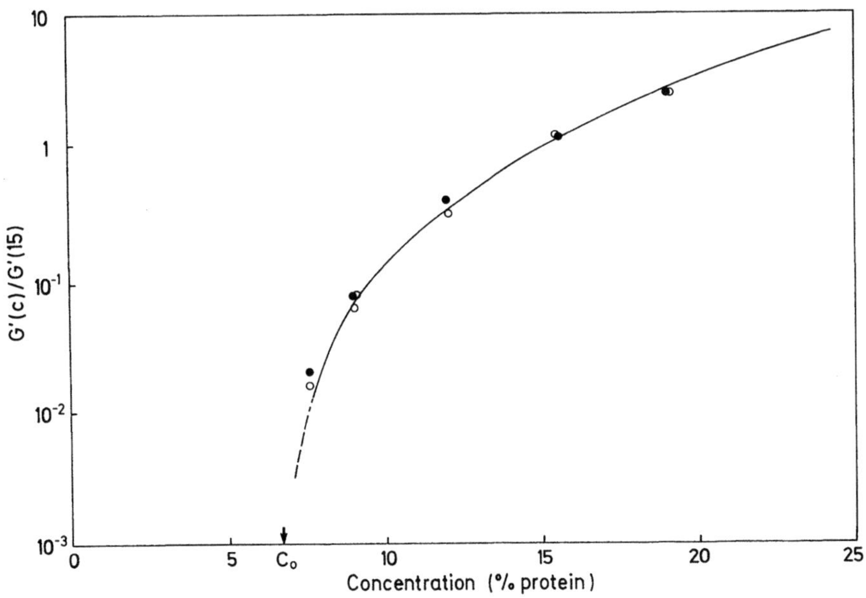

Fig. 63. G′ (●) and G″ (○) for BSA gels (pH 6.3) scaled to unity at concentration C = 15% w/w, and plotted against concentration. Solid line is theoretical function of Hermans. Reproduced with Permission from British Polymer Journal, *13*, 11 (1981), Fig. 2, p. 12

relationship for biopolymer gels in a "universal" way, using cascade theory. Further information about this development may be obtained in Section 3.4 and in the review by Clark and Lee-Tuffnell [35] where the underlying assumptions of the modified Hermans treatment, and its adequacy to describe biopolymer gel data, have been critically assessed.

Turning to the behaviour of heat-set globular protein gels at larger deformations, the strain dependences of G′ and G″ for BSA gels have already been referred to, and a linear viscoelastic region extending to strains of at least 0.10 remarked upon. Above this level of strain, the modulus may change somewhat, but the stress-strain profile will still be that of a solid. Rupture of the gel will eventually occur, however, probably at fairly low strain, as heat-set protein gels are usually brittle.

Detailed experimental information about the large deformation behaviour of heat-set globular protein gels is quite sparse, but some data can be found. Hermansson, for example, has made a study [347] of the large deformation and failure properties of gels based on blood plasma protein (mainly BSA) and has used penetration and compression tests. From the initial slopes of force-versus-displacement curves, estimates of gel rigidities were obtained for gels made under a variety of conditions. In general it was found that the most turbid gels were the most brittle, and trends in the rigidity values in relation to changing gelling conditions were broadly in agreement with trends suggested by Richardson and Ross-Murphy. Such differences as did emerge could be explained in terms of the high salt contents of the blood protein samples examined in Hermansson's study.

5.3.2 Gels from Chemically-Denatured Globular Proteins

Globular proteins can be denatured by agents other than heat [348], and when this happens, association of the chemically unfolded species can take place by a branching mechanism, much as in the heat-setting situation. Alcohols, for example, can produce gels in this way [349], and a combination of urea and heat can cause ovalbumin solutions to form partially covalent networks [350]. None of these systems has been well characterised structurally, but an interesting linear viscoelastic investigation of the ovalbumin/urea gel has been published [350].

Branched, aggregated, structures involving globular proteins can also be induced by other non-thermal procedures. Antibodies, for example, can cross-link antigen proteins [34] into a three-dimensional network, much as red blood cells can be cross-linked in blood agglutination. The aggregation of casein micelles, to be discussed in the next Section, is a process induced by enzyme catalysed removal of peptide chains from parts of the casein micelle surface.

5.4 Casein Aggregation and Gelation

5.4.1 Introduction

Casein is the principal protein component of milk, and is a highly polydisperse substance consisting of roughly spherical molecular aggregates (micelles) varying in diameter from ~ 20 nm to ~ 300 nm [351–354]. The weight average molecular weight of these particles is large, being of the order of 3×10^8 and their high voluminosities [352] ($\sim 4\ m^2/gm$), and high water to dry protein weight ratios, suggest open spongy structures. In fact, casein micelles have been shown to be composed of very large numbers (up to thousands) of sub-micellar proteinaceous particles connected into open network arrangements [351] the integrity of which is belived to be maintained by inorganic material such as colloidal calcium phosphate (Fig. 64) also present. The submicelles have been characterised structurally by neutron scattering [353] which suggests a weight average molecular weight of 3×10^5 and an average radius of gyration of ~ 6.5 nm but since, currently, their existence as discrete entities in the casein micelle has been challenged [354] in favour of a fused filamentous model, this description should perhaps be regarded as provisional.

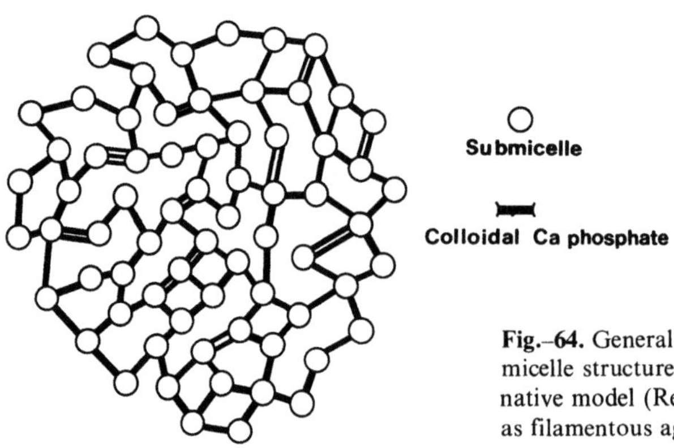

Fig.–64. Generally accepted representation of casein micelle structure. Note that in a more recent alternative model (Ref. [354]) the sub-micelles are present as filamentous aggregates

At a molecular level the sub-micelles are substantially heterogeneous being composed of more than one molecular protein component. Depending on sequence and molecular weight, these individual protein constituents are described as α_s (actually α_{s_1}, α_{s_2}), β and $\varkappa$ caseins, and they are usually present in sub-micelles in the proportions 4:3:1. Their individual molecular weights lie in the range 19,000 to 25,000, and special structural features include the presence of hydrophobic residues in regions along the peptide chains, ester phosphate groups attached at certain points, and a distinct lack of ordered secondary structure. Features such as these confer on these molecules the capacities (to varying degrees) to act as detergent-like protein species and to bind cations (e.g. Ca^{++}), and these properties are clearly utilised and manifested in the constitution and behaviour of casein in all its forms.

Although present in caseins to the smallest extent ($\sim 12\%$ w/w), $\varkappa$-casein is a highly significant constituent, as it is known to promote stability of the micelles against aggregation. $\varkappa$-casein is believed to be particularly prevalent on the micelle surface and its stabilising influence may be due to its provision of 'hairy' hydrophobic glycopeptide protrusions [352] which give rise to entropic repulsion. Certainly, removal of these 'hairs' by proteolytic enzymes leads to a dramatic loss of casein micelle stability, and aggregation [355–358], and even gelation, can take place (as, for example, in cheese-making).

Whilst enzyme action is one method of inducing casein solutions to aggregate or gel, other mechanisms for inducing attraction exist. These include lowering pH (as in the manufacture of yoghurts [354, 359]), adding specific ions such as calcium ions, heating casein solutions, or applying some combination of these.

In view, however, of the considerable practical importance of enzyme induced gelation of bovine milk casein in cheese manufacture, enzyme-induced aggregation of casein will be discussed first. Gelation of casein by other mechanisms will be considered subsequently.

5.4.2 The Enzyme-Induced Aggregation of Casein Micelles

5.4.2.1 Introduction

The action of rennet (mainly the enzyme chymosin) on the casein component of milk forms the basis of cheese manufacture. This process leads to a gel, or curd, which is eventually cut and, after drainage of the whey, is allowed to mature.

Formation of the gel (or indeed aggregates of any size) requires time, and the time lag before clots become visible has loosely been described as the rennet coagulation time (RCT). The action of rennet on diluted casein preparations has usually been followed by turbidity measurements, and the coagulation time then estimated from plots of turbidity against time (Fig. 65). The turbidity approach is now examined in more detail.

5.4.2.2 Turbidity Measurements

Studies of enzyme-induced casein aggregation by turbidity measurements have been pursued for many years, and these have been accompanied by a growing body of theory aimed at both quantitatively and qualitatively describing the lag phase and its

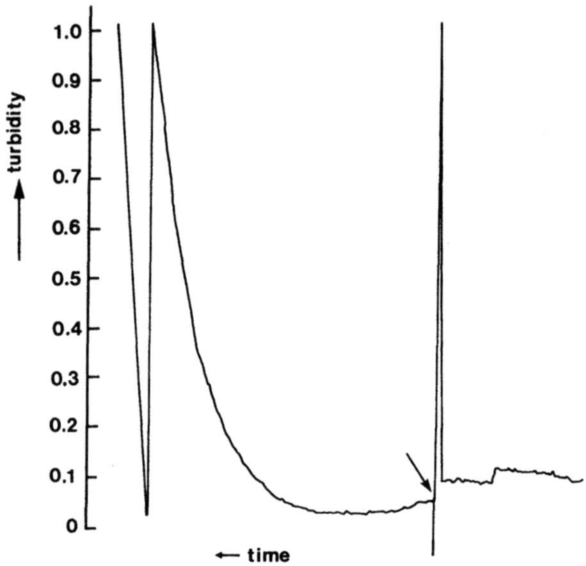

Fig. 65. Change in measured turbidity with time as casein micelles in solution aggregate as a result of enzyme (rennet) action. Arrow indicates time at which enzyme was added. Initially, turbidity falls slightly (see text) before increasing substantially as aggregation takes place. A coagulation time can be estimated by extrapolating the steep part of the turbidity-time curve to meet the time axis. Reproduced with Permission from Faraday Discuss. Chem. Soc. *57*, 164 (1978), Fig. 2a, p. 166

relationship to enzyme concentration. An explanation of the influence of changing casein concentration has also been sought.

Theoretical description of molecular weight changes in aggregating casein solutions, and interpretation of the lag phase, has been pioneered by Payens [355–358]. Payens' approach is based on combining Michaelis-Menten kinetics to describe the enzyme action on the ϰ-casein in the micelles (splitting off of the C-terminal glycopeptide) with Smoluchowski's [360] theory of particle aggregation to describe the subsequent association of activated casein (paracasein) species.

In his original treatment of the problem [355–357], Payens made the simplifying assumption that the rate of production of aggregable paracasein micelles was proportional to the rate of cleavage of ϰ-casein molecules on the micelle surfaces, and that throughout the time interval of interest, this rate was constant. Later, however, these conditions were relaxed in that the rate of production of paracasein was allowed to vary [358]. Payens still retained the principle, however, even in this more general approach, that the rate of production of active casein was proportional to the rate of ϰ-casein cleavage. The more general treatment allowed the weight average molecular weight (M_w) of an aggregating casein solution to be written as,

$$\left[\frac{M_w}{M_0}\right]_{\text{total}} = 1 - M_0(1-r)\,P(t)\left\{2r - (1-r)\,k_s\,\frac{\int_0^t P(t)^2\,dt}{P(t)}\right\}(C^0)^{-1} \quad (36)$$

Here P(t) is the total number of active paracasein species produced at time t, C° is the initial concentration of the casein, M_0 is the substrate molecular weight, r is the ratio of the molecular weight of the cleaved peptide to that of the substrate, and k_s is the Smoluchowski rate constant for the association process.

Eq. (36), derived by Payens, is in fact Smoluchowski's original theory, generalised to include an arbitrary production of reactive monomers during the course of the aggregation process. Smoluchowski's original linear increase of molecular weight with time (and no lag period) is recovered if P(t) is constant. If, however, as was originally assumed by Payens in his more restricted model [357], active species are produced at a constant rate V_{max} (the maximum rate of enzyme cleavage predicted by the Michaelis-Menten equation, and a rate proportional to enzyme concentration) the above equation may be re-written as,

$$\left[\frac{M_w}{M_0}\right]_{total} = 1 - M_0(1 - r)\,(8V_{max}/k_s)^{1/2}\,\{r(t/\tau) - (1 - r)\,(t/\tau)^3/3\}\,(C^0)^{-1} \tag{37}$$

which now contains the time, scaled by a quantity τ given by,

$$\tau = \sqrt{\frac{2}{k_s V_{max}}} \tag{38}$$

This second equation for the overall weight average molecular weight of the system predicts that this quantity will fall at first, pass through a minimum, then rise sharply at times greater than τ. The slight fall in molecular weight, which has been detected in practice by viscosity studies, is a result of the casein molecular weight initially being reduced on account of the enzyme action. The aggregation process eventually counters this effect and M_w starts to increase again.

In applications of Eq. (37) for M_w, Payens regarded τ as the coagulation time (though experimental measurement by extrapolation procedures applied to turbidity curves were, as he concluded, likely to generate a quantity proportional to τ) and he remarked upon the equivalence of k_s and V_{max} in their determination of τ. He also noted that, at constant k_s, τ would vary as $V_{max}^{-1/2}$, or as the enzyme concentration to this same power. The fact that, in practice, a reciprocal relationship between τ and enzyme concentration, is usually observed, was explained by the assertion that k_s itself might be influenced by enzyme concentration, and hence depend on V_{max}.

Whilst the Payens model is able to explain the basic features of enzyme induced aggregation of casein, and is reasonably successful, quite recently some of its basic assumptions have been challenged. A major issue has been the relationship between the rate of enzyme action, and the rate of aggregable casein production. Dalgleish, for example, has employed an ingenious experimental approach [361] to investigate this aspect of the problem, and has concluded that, in practice, the micelles show little tendency to aggregate until at least 88% of their surface κ-casein has been cleaved. In contradiction to Payens original assumption, therefore, Dalgleish's findings suggest a highly non-linear relationship between enzyme reaction extent and the number of reactive paracasein molecules produced, and on this basis he constructed a new mathematical model [361,362].

The Dalgleish model is also based on the generalised Smoluchowski equation (Eq. (36)) and on Michaelis-Menten kinetics, but it invokes a more complex relationship between P(t) and the extent of enzyme reaction at time t, than was previously adopted by Payens. The new model recognises a cleavage threshold for any casein micelle, below which it is unreactive. Coupled with an assumption of random enzyme attack, and an experimentally determined threshold value, P(t) is calculated from such information as the substrate concentration, the maximum rate (and hence the enzyme concentration) V_{max}, and the enzyme velocity constant. From P(t), $M_w(t)$ is then calculated numerically using Eq. (36).

In Dalgleish's approach, the rennet coagulation time (RCT) cannot be formulated analytically, but its dependence on variables such as the enzyme concentration can still be explored. On the assumption that the RCT could be defined as the time required for M_w to reach some constant multiple of the original casein monomeric molecular weight, Dalgleish [362, 363] was able to demonstrate numerically that his model was consistent with the expected reciprocal relationship between RCT and enzyme concentration. There was also the additional bonus that the model successfully predicted that the RCT would show little dependence on substrate concentration until very low concentrations were reached [363]. In this latter connection, it is interesting to note that Darling and co-workers [364], using a somewhat different Smoluchowski approach to casein aggregation, have also been able to explain the RCT-substrate concentration dependence (and indeed, the enzyme concentration dependence as well). Their model assumes a constant number of reactive particles in the aggregating casein system, and focusses attention on a changing Smoluchowski rate constant for these particles. The Dalgleish analysis may accord more closely with molecular events, however.

When applied to turbidity data, both Payens' and Dalgleish's approaches allow rate constants k_s to be determined. Studies of the value of these constants by both authors [351, 357, 365] have cast light on the nature of the casein aggregation process. It was concluded early on, for example, that values obtained for k_s, particularly at ambient temperature, were orders of magnitude smaller than would be expected for a diffusion controlled process, and studies of the variations of k_s with temperature and ionic strength made it clear that a simple electrostatic barrier was not the explanation for the low reaction rate. Both Payens and Dalgleish concluded that classical theories of colloid stability invoking balance between repulsion and attraction (DLVO theory) are not sufficiently sophisticated to treat the casein situation. Specific interactions based on both hydrophobic and ionic interactions are likely to be involved in casein aggregation, and reactive sites [351] ('hot spots') may be present at various points on micelle surfaces (Payens). This last proposal, put forward by Payens [351], but not supported by Dalgleish on account of his findings that large regions of the micelle surface must be denuded of ϰ-casein before aggregation can begin, implies discrete cross-linking functionalities for casein micelles, and this in turn suggests that a branching process may be involved in casein aggregation. The turbidity studies just described were performed at low casein concentrations, and whilst in these circumstances, Smoluchowski kinetics may be applicable, casein systems at higher concentrations can gel, and this must favour the branching hypothesis. Gel formation is now considered.

5.4.2.3 Rheological Studies

Early studies of casein aggregation involved viscosity measurements, and viscoelastic characterisation using simple U-tube gelometers. More recently Culioli and Sherman, have also used apparatus of this kind (Scott-Blair and Burnett, and Saunders-Ward[151]) in their studies[366] of the time courses of modulus changes during casein gelation at various protein concentrations (~ 3 to 20% w/w). In Culioli and Sherman's work experiments were also performed for various concentrations of added chymosin, and fully cured gels were examined by performing creep and creep recovery experiments.

The studies of Culioli and Sherman[366] led to some interesting conclusions. First, the times required for the moduli of given casein gel systems to become measurable (gel times) were identifiable with visually observed coagulation times (RCT's). Second, for any system, the concentration of casein was found to determine the ultimate (long-time) value of G'. Third, for a given protein concentration, decreasing the enzyme concentration lengthened the gel time, and produced a much slower growth of G' with time. Fourth, for a given enzyme concentration, the gel time was lengthened by increasing protein concentration, though this increase was less marked at high enzyme concentrations, and fifth, the plateau G' values showed a $C^{2.6}$ concentration dependence. In general Culioli and Sherman's experiments indicate linear viscoelastic behaviour, and their power law for the modulus is in line with results for other biopolymer systems (cf. Sect. 3.4). Their gel time observations in relation to protein concentrations, however, are perhaps less expected in the light of Dalgleish's results for substrate concentration effects, discussed earlier.

The kinetic development of shear modulus in casein-rennet systems has also been studied by Johnston[120]. In this work, Flory's gel theory[110], combined with the theory of rubber elasticity[78], was used to construct a model to fit the time course of modulus development. The model relied on three parameters: (1) a rate constant for the second-order cross-linking of reactive sites: (2) the total number of sites in the system available to from cross-links, and (3) the total number of units in the system involved in cross-linking. Using this model, sets of modulus-versus-time data were fitted, and the parameters optimised by least squares. Shifts in parameter values were monitored in relation to changing gelation conditions such as calcium ion concentration, and in this last case the rate constant for cross-linking, was found to increase as the ion concentration increased. A C^2 dependence for the gel modulus was also reported, as was a linear relationship between the breaking force of the gels (cone penetrometer) and their moduli. In a later paper[367], Johnston also reported stress relaxation measurements for casein gels, and from the fact that complete relaxation occurred he concluded that physical forces predominate in maintaining network cohesion. Interestingly, Johnston found two distinct regions in the stress-relaxation decay, but no molecular explanation was given for this.

It is worth noting that Johnston's approach has recently been criticised by Payens[368] and by van Vliet and Walstra[369], these last authors taking particular exception to the use of ideal rubber theory in the casein gel context, and by implication, to an entropic description of the free energy increase involved in deforming casein networks. In view of the apparently corpuscular nature of casein gel structures, this objection seems reasonable even though, in reality, the existence of a strong enthalpic contribution to the casein gel modulus has not yet been established unequivocably. If necessary, however, Johnston's approach could be generalised to include a non-ideal front factor as described previously in Section 3.4, though this would be at the expense of introducing another variable parameter.

Another recent rheological study of enzyme-induced casein gels is that by Tokita and co-workers [62]. Here, the concentration and frequency dependence of the gel modulus, close to the critical concentration, was examined. Accurate modulus measurements at known frequencies and strain, using a torsion pendulum, provided data which was interpreted within the framework of percolation theory [125]. This data related to casein systems studied at concentrations in the range 0.60, to 2.35 gm/100 ml, the enzyme-substrate ratio being maintained at 4.2×10^{-3}. G' was evaluated over a frequency range of 1.0 to 5.0×10^{-3} Hz and at a strain (1.4%) small enough for a linear viscoelastic response to be assumed. The temperature was 20 °C ± 0.2 °C, and the dynamic shear modulus was recorded only after extensive cross-linking (~5 hours) had occurred. Inspection of G' versus concentration data revealed a critical concentration for gelation of $C_0 = 0.56$ gm/100 ml, and below this limit only aggregates formed. Modulus concentration plots were established at each frequency, and data fitted to a relationship

$$G = C'\varepsilon^q \tag{39}$$

with

$$\varepsilon = (C - C_0)/C_0 \tag{40}$$

q and C' being determined by least-squares. The values obtained for C' and q varied from 12.7 to 15.8 for C' and from 2.6_9 to 1.8_2 for q, the highest q result being obtained at 1 Hz. Below 1.0×10^{-1} Hz an approximately constant value for q of 1.8–1.9 was found. Tokita et al. noted that percolation on a 3-D lattice graph would be expected to give a critical exponent $q = 1.88$ and claimed excellent agreement between their experimental results and percolation theory predictions.

In their work, Tokita et al. made no reference to possible fits to modulus concentration data using other approaches (for example, classical branching theory). This omission has been commented upon by Gordon [115], who has considered a number of classical 'so-called' mean-field gelation models in relation to the casein data, and has shown how all of these can be fitted to the experimental data with comparable success (Fig. 66). The models chosen were based on Flory's random f-functional polycondensation approach, and differed in terms of the unit (whole micelle, submicelle, polypeptide chain) assumed to be the aggregating species. Reversibility of bond formation was assumed, and the extinction probability concept, and mathematical framework of the cascade theory of branching, used to calculate gel moduli. This approach is in fact similar to that described by Clark and Ross-Murphy [70] to calculate biopolymer gel moduli at concentrations considerably higher than critical but Gordon, in his calculations, (like Johnston), stuck rigorously to the ideal rubber elastically active chain contribution to the modulus, kT. Statistical analysis of fits to the casein data of Tokita and co-workers corresponding to the various Gordon models showed clearly how indistinguishable they were, not only in relation to each other, but also in relation to the fit allowed by the percolation approach.

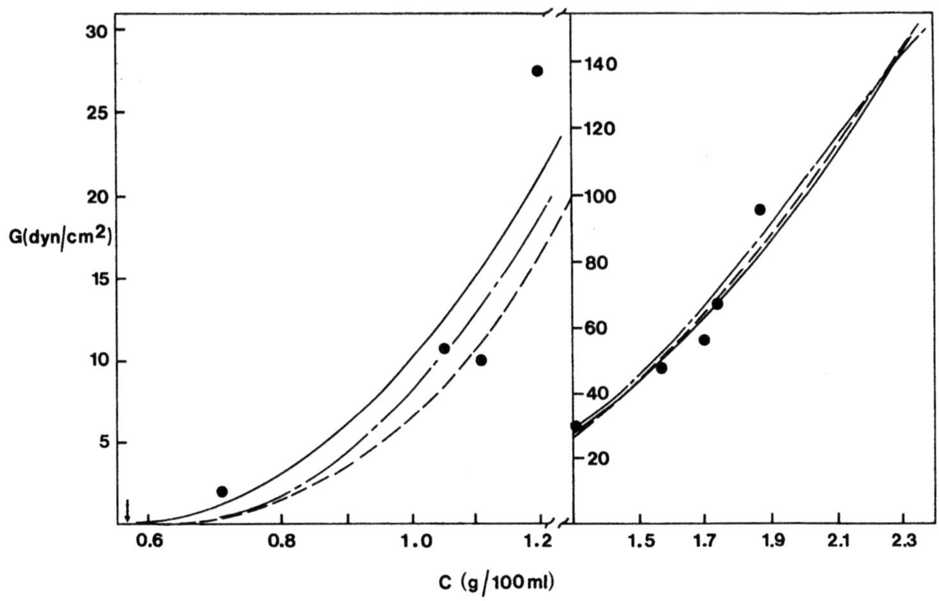

Fig. 66. Fit by Gordon (Ref. [115]) to the data of Tokita and coworkers (Ref. [62]) for casein gels. Experimental points (●); curves are: for percolation model-solid line; for classical vulcanisation model for submicelles-dashed-dot line; for classical random f-functional for submicelles-dashed line. Reproduced with Permission from "Integration of Fundamental Polymer Science and Technology", edited by L. A. Kleintjens and P. J. Lemstra, Elsevier Applied Science, 1986, Fig. 1, p. 169

5.4.3 Other Forms of Casein Aggregation

5.4.3.1 Thermally-Induced Aggregation

This is a well-established phenomenon for whole milk which occurs at temperatures of ~140 °C. A lag phase is observed, followed by a fast casein precipitation, and several years ago, when studying this event, Parker and Dalgleish [370] proposed that classical branching theory (in particular, the cascade branching model [112]) could be used to describe its time course. According to Parker and Dalgleish, the polyfunctional condensation approach could explain experimental measures of lag times, and the concentration dependence of these. Where more complex behaviour was encountered, as happened for some milks, non-equivalent functionalities were proposed.

5.4.3.2 Calcium-Induced Aggregation

Both micelles and casein protein components, derived from micelles, can be aggregated by calcium ions. Interaction between α_{s_1}-casein and calcium has been most thoroughly investigated. The α_{s_1}-casein carries a negative charge, and calcium binding reduces this, generating instability. A lag period is observed, followed by a period of rapid rise in molecular weight and increased turbidity. Light scattering experiments [371], using a stopped-flow method, provided molecular weight information, and Parker and Dalgleish proposed a branching description to explain the lag period, but a Smoluchowski model to explain later aggregation events. Recently, however, as a result of more extensive studies of the phenomenon, Dalgleish and co-workers [372] now ascribe the lag phase to the time needed for α_{s_1}-casein monomers to form octamers. The octamers then, it is proposed, aggregate via Smoluchowski kinetics.

In casein aggregation, it appears, the need for branching theories, and for non-Smoluchowski kinetics, seem only to survive for the heat-induced aggregation process, and for rennet-induced gelation at higher concentrations. Other situations seem to be satisfactorily described by the Smoluchowski model, when this is coupled with some other rate limiting step controlling the supply of aggregable particles.

5.4.3.3 Aggregation on Lowering pH — Acid Casein Gels

Aggregation (and, under certain conditions, gelation) of casein in acid conditions produced by bacterial action is the basis of yoghurt-making [354, 359]. Originally this phenomenon was believed to involve the coming together of casein micelles under conditions of decreasing micelle charge (casein isoelectric point is at pH $\sim$ 4.6) but recent research has shown [354] that the process is more complicated than this and involves fundamental changes in the original micelle structure. Thus, as the pH falls from its original value in milk (greater than 6.0) the inorganic material (i.e. amorphous calcium phosphate) within the micelle structure dissolves out, and some of the protein components also become solubilised. As the pH continues to fall, however, a complex sequence of events take place ending in reaggregation of the protein part of the system and the formation of new protein particles flocculated into a coarse-stranded particulate network.

Detailed rheological studies by Roeff [359] of acid casein gels induced by acidification in the cold ($\sim$4 °C) followed by moderate heating (i.e. acid casein gels similar to but not quite identical to the yoghurt-like preparations of Ref. [354] have demonstrated their non-equilibrium character (G′ still varying with time at long times), and accompanying microscopical examination (transmission electron microscope on sections) coupled with permeability studies have established their particulate and heterogenous microstructures. A study of the concentration dependence of the gel modulus (gels cured for many hours) has suggested a $G' \sim C^{2.6}$ power law much as was found by Culioli and Sherman [366] for the rennet-induced materials, and transient shear modulus-time experiments have indicated the presence of both long ($>10^4$ sec) and shorter (60 to 6.0×10^3 sec) lived crosslinks.

The acid casein gels are clearly more complex structures than the rennet gels discussed earlier, but at the same time they have both structural and rheological features in common with these. The complicated pathways by which they form, however, and the crucial influence on their final properties of factors such as pH, salts, protein concentration and thermal history, make them likely to be more difficult subjects for systematic study, particularly if quantitative theories of network formation are to be profitably applied.

5.5 Networks from Rod-Like Polypeptides and Proteins

5.5.1 Helical Polypeptides

5.5.1.1 Introduction

It has long been known [373, 374] that solutions of helical polypeptides such as poly(α-benzyl-α, L-glutamate) PBLG and poly(ε-carbo-benzoxyl-α, L-lysine) PCBL show complex phase behaviour, including the formation of gels. Both disordered and liquid crystalline phases can occur, and phase diagrams indicate conditions of temperature and composition under which these phases co-exist, or are individually stable.

In general, results for these systems qualitatively confirm Flory's theoretical phase diagram [146] for rigid rods in solution, (Fig. 67). Since polymers such as the polypeptides just mentioned, are known to adopt rod-like α-helical conformations in the solvents ('helicogenic') used in experiments (N,N-dimethyl formamide, benzene and toluene), there is a clear vindication of Flory's approach. Such detailed differences between theory and experiment as have been found can realistically be explained in terms of incomplete peptide rigidity, and a limited permeability to solvent [303].

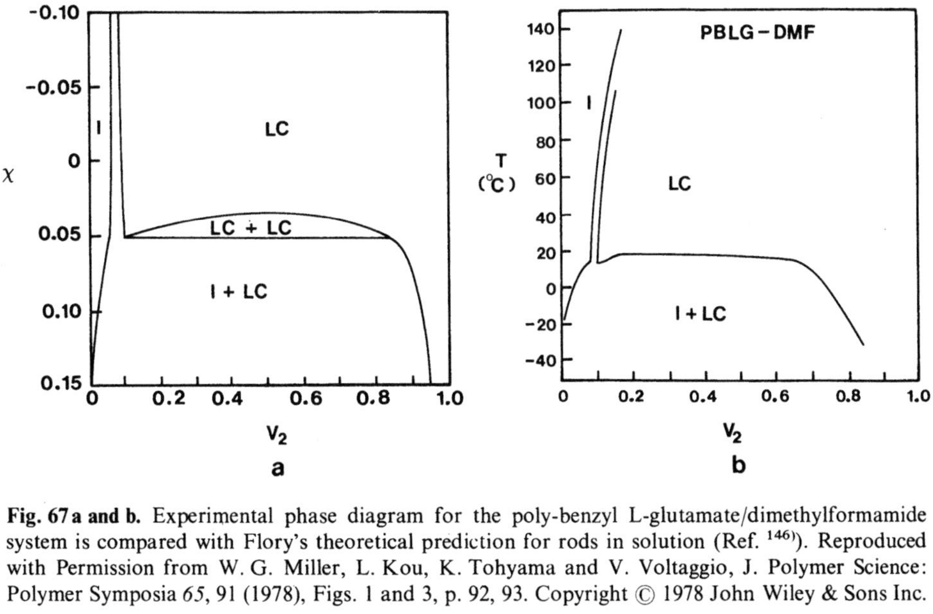

Fig. 67 a and b. Experimental phase diagram for the poly-benzyl L-glutamate/dimethylformamide system is compared with Flory's theoretical prediction for rods in solution (Ref. [146]). Reproduced with Permission from W. G. Miller, L. Kou, K. Tohyama and V. Voltaggio, J. Polymer Science: Polymer Symposia *65*, 91 (1978), Figs. 1 and 3, p. 92, 93. Copyright © 1978 John Wiley & Sons Inc.

A typical experimental phase diagram [303] for the PBLG-dimethyl formamide system also appears in Fig. 67, and shows how at higher temperatures, and within a narrow composition range, two phases co-exist, a disordered solution and a liquid crystalline phase. Below this so-called 'chimney' region there is a broad biphasic area covering a wide composition range, and again co-existence of disordered and liquid crystalline phases is indicated. At lower temperatures the compositions of the co-existing phases may be seen to tend towards pure solvent and pure polymer, and a particularly interesting feature of this biphasic state is that it usually shows the mechanical characteristics of a gel. This gel state, which implies a three-dimensional network, has long been recognised, and recent research has focussed on structural and rheological characteristics of the network involved [303].

5.5.1.2 Structural Aspects

Miller and co-workers have made extensive studies of the phase behaviours of rod-like polypeptides in non-aqueous solvents [303]. They have monitored the formation, and ageing, of the liquid crystalline phase, for example, and have characterised the biphasic gel state. This latter they have described as a transparent, mechanically self-supporting, material, whose formation is concentration and temperature dependent, and which is completely reversible. They report that gels can be formed at concentrations as low as 0.03% w/w polymer and they point out that, in the networks implied, there is no likelihood of covalent cross-links, or even of hydrogen bonds. A model involving polymer chains trapped between crystalline regions can also be excluded, leaving only a model of rod-like particles aligned into network strands, or sheets, many molecules thick. This description has been substantiated by microscopy studies, and Tohyama and Miller [375] have compared and contrasted the networks produced with those of gelatin gels observed by the same (freeze-fracture) electron microscope approach. In the micrographs discussed, the gel networks are apparently composed of strands or fibres that appear to merge into one another rather than cross-link at specific points. The strand thicknesses vary from tens to more than one hundred nanometers (the individual rod has a diameter of <2.0 nm), and the spacing between fibres is typically 1 μ.

Miller and co-workers [302, 303, 375] explain the formation of the structures just described as a kinetic event, realising the natural thermodynamic drive towards phase separation, but occurring without activation, and without the involvement of nucleation and growth steps. It is argued in this case that network formation is a consequence of the Cahn-Hilliard mechanism of spinodal decomposition (see also Sect. 3.7), which predicts phase separation by a non-nucleated mechanism [150, 376], when a solution is brought into a region of thermodynamic instability ($\delta^2G/\delta C^2 < 0$, where G is the gels free energy and C the polymer concentration). This theory predicts continuity for both separating phases, early on in the phase separation process, and anticipates a periodic structure, with a characteristic wavelength. Although such a structure is inherently unstable, and should 'age'; in the helical peptide gels the rigidity of the peptide phase inhibits this. The characteristic 'spacing' of the network, which is approximately a micron, is understandable (so Miller proposes) in the light of the Cahn-Hilliard analysis.

5.5.1.3 Rheological Properties

Miller and co-workers have also studied the mechanical properties of these gels [303]. Mechanical measurements on a 1% w/w PBLG solution, in toluene, at about 25 °C were made using a Rheometrics Rheometer, and the dynamic shear modulus was determined as a function of frequency (0.04 to ~ 0.16 Hz) and rod concentration. The storage modulus G′ was found to be $\sim 1.0 \times 10^4$ dyne/cm^2, and the loss tangent was small. As far as could be judged, G′ showed little frequency dependence, and perhaps, much more surprisingly, little concentration dependence. Later studies of PBLG, in toluene and N,N-dimethylformamide and over a 300 fold range in concentration, found only a three fold change in G′, and the relationship, $G' \propto C^{0.2}$, which this finding suggested, seemed considerably at odds with the C^2 dependence often quoted for bipolymer gels (Sect. 3.4). Variation of G′ with temperature showed no particular trend, and showed no tendency to follow the ideal rubber theory prediction of strict temperature proportionality.

5.5.1.4 Conclusions

The known facts about the gels discussed in this Section have allowed conventional network descriptions for them to be rejected, particularly including Doi and Kuzuu's 'brushpile' model [105] for rods. Instead a network state, generated kinetically by spinodal decomposition, seems indicated, although it is still largely unclear why this mechanism should produce gels whose moduli vary so little with concentration (cf. Sect. 3.7).

The spinodal decomposition mechanism may, as Miller and co-workers, have suggested, be followed by other gelling biopolymer systems. Under certain conditions, rod-like molecules such as actin and tubulin (see earlier discussion) might be expected to behave in this way, and Miller has suggested that the Hb-S polymer, also discussed previously, forms this type of network. The tobacco mosaic virus, a rod of 150 nm by ~3000 nm is known to show complex solution behaviour [3] including gel formation, but so far, no data is available about the network structure, or its mechanical properties. Clearly it remains to be established how common spinodal decomposition of rod-like molecules really is as a gelation mechanism for biopolymers in aqueous media.

5.5.2 Myosin

5.5.2.1 Introduction

Myosin [284] is the principal protein component of muscle. It is a rod-like molecule composed of two equivalent high molecular weight (~200,000) peptide chains, but deviates from rod geometry in that the extended chains develop into two globular regions at one end as is shown schematically in Fig. 68. The myosin molecule is also believed to have associated with it four much lower molecular weight peptides, non-covalently bound to the globular headgroups. In outline, therefore monomeric myosin can be described as consisting of two parts, a headgroup region and a tail, the full length being ~150 nm, the tail cross-section ~2.0 nm, and the overall molecular weight ~500,000. In the tail region the two peptide chain components take up almost purely α-helical conformations, whilst in the headgroups they develop compact globular forms, with only ~30% of their peptide residues in the α-helical state. Another important feature of the tail is the presence of a so-called 'hinge' region (see also Fig. 68) which is sensitive to attack by the proteolytic enzyme trypsin, and which confers upon the molecule a degree of flexibility necessary for its biological function.

Although the myosin molecule is an important, indeed essential, component in determining the contractile function of muscle, it is only one part of this highly

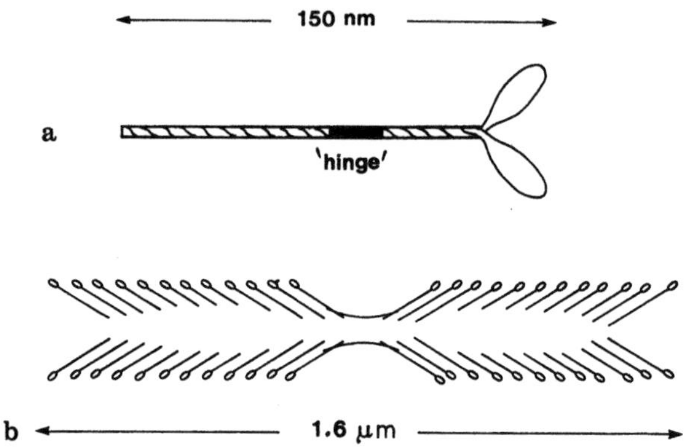

Fig. 68 a and b. Representation of the myosin rod structure (**a**) and of the mode of myosin aggregation to form the thick filaments of muscle fibres (**b**)

complex protein-based mechanism, other proteins such as actin (Sect. 5.2.1) tropomyosin, and troponin being involved [284]. In the myofibrils, or muscle cells, ultimately composing whole muscle, thick and thin filaments, consisting of these molecules, form interpenetrating structures capable of lengthwise contraction by what amounts to a ratchet mechanism. The thin filaments contain chains of actin molecules attached to tropomyosin threads, whilst the thick filaments are bipolar aggregates of myosin molecules (see also Fig. 68). During muscle contraction, myosin head-groups, present on the surfaces of thick filaments interact with actin in the thin filaments, this process being influenced by calcium ions, and by the biochemically important energy source adenosine triphosphate (ATP). In the gelation of myosin extracted from muscle, both the tendency for myosin to self-associate into filaments, and to interact with actin are important.

5.5.2.2 Myosin Gels

The gel-forming behaviour of myosins from various muscle types has been extensively studied [377–383]. Occasionally such gels can occur spontaneously at ambient (or lower) temperatures, but more usually heat-setting is required, as in the case of the gelling of globular proteins discussed earlier. Heating of myosin solutions to at least 60 °C is necessary, and the resulting gels depend crucially for their properties on gelling conditions of pH and ionic strength. Myosin monomers (i.e. the two-chain species referred to in the Introduction) can only exist free in solution (or as very small aggregates) under conditions of high pH (typically 6 to 8) and high ionic strength. As ionic strength falls, at pH 6.0 for example myosin molecules spontaneously re-associate to form structures analogous to the muscle thick filaments, and it follows that gels formed under these conditions will be based on fundamentally different underlying units, than those arising from the monomeric solutions [384]. The low ionic strength gels will henceforth be described as filamentous.

Non-filamentous myosin gels can be made at pH 6.0 and in 0.6M KCl, by heating such solutions above 60 °C, and in these circumstances strong elastic gels are produced. The nature of this process has been studied by Samejima and co-workers [385] using specially prepared head and tail fractions from an original myosin preparation. This work has suggested that, whilst network formation from monomeric myosin is mainly determined by thermal denaturation of the helical tail sections of myosin molecules at 54 °C, and the subsequent aggregation, a preliminary aggregation step takes place at much lower temperatures, and this involves the head-groups. The tail-tail interactions are primarily non-covalent, but disulphide cross-links are implicated in the head-group cross-linking process. Electron microscopy [384] has suggested that networks formed from monomeric myosin, at high ionic strength, are substantially heterogeneous.

Gels formed from myosin filaments are rather different. The filamentous solutions/suspensions formed at low ionic strength contain very large aggregates (several hundred monomers). Heating the resulting solutions/suspensions, however, produces a very strong elastic gel, and recent scanning electron microscopy of such materials, indicates a uniform microstructure [384]. These gels are apparently composed of a uniform distribution of large filamentous species cross-linked via the head-group sites distributed over the filament surfaces. This description seems supported by the fact that shear modulus-versus-temperature profiles [386] for the curing of filamentous gels show a growth of G' prior to tail-section unfolding, then a sharp drop in G', as tail unfolding takes place, followed by a continued increase in the modulus after this point. Thus, whilst the 'melting' of the helical core of the filaments makes itself felt during the curing process, cross-linking is established much earlier, via the surface head-groups. Similar cure curves for the high ionic strength, monomeric, myosin gels apparently show no signs of the modulus drop. In terms of the influence of concentration on G' it is of interest that power laws of the form $G' \propto C^n$ have been obtained for various myosin gels with the exponent n lying in the range 1.8 to 2.0 [387]. This exponent seems to provide clear evidence that myosin gels are in no way similar to the helical polypeptide gels discussed by Miller. They have much more in common it seems with the gels formed from denatured globular proteins discussed in Sect. 5.3.1.

Finally, it should be added that actin-myosin mixed gels have not been included in this Section as these are essentially gel composites. Suffice it to say that, from their molecular relationship in muscle cells, it is hardly surprising that myosin can affect actin network formation, and *vice versa*.

5.5.3 Collagen

The essential subunit of collagen, tropocollagen [151, 152], is another example of a semi-rigid rod-like biopolymer. Collagen itself is a complex fibrous, proteinaceous material, which is the principal protein component of white fibrous connective tissue. It is found in skin, tendon and bone and indeed in almost every part of mammalian tissue. Although most collagen is water insoluble, some fractions of collagen from mammalian sources, are soluble in acidic buffers, and can be persuaded to yield solutions of a substance of much lower molecular weight, tropocollagen.

Tropocollagen, as has been shown by light scattering [152], is a rod-like particle (Fig. 23) of molecular weight ~345,000, length 300 nm and diameter ~1.4 nm. At concentrations much above 2–3 mg/ml aggregation occurs yielding viscous solutions, and dialysis of acidic solutions tends to produce re-assembly of the original collagen fibres. A specific head-to-tail aggregation process is involved accompanied by lateral staggering, and aggregates have a characteristic striation. It is clear that this is a much more specific form of interaction than that leading to helical polypeptide association, and does not appear to be analogous to it. The specific nature of the interaction depends on the peculiar structure of tropocollagen which as was mentioned earlier (Sect. 4.1) is a triple helix (Fig. 23) involving three proline rich peptide chains. The particular peptide sequences of these chains give rise to areas on the rod surface prone to ionic and hydrophobic interaction and this situation seems to produce the directed aggregation observed [151].

Heating tropocollagen solutions leads to co-operative destruction of the triple helix and separation of the chains. The resulting solution contains gelatin (the denatured product of collagen) and, as has been discussed at length in Sect. 4.1, gelatin solutions gel on cooling rather than renaturing to the original collagen state. A process of renaturation still occurs, but this is frustrated by the intrinsic disorder of the solution situation, and only partial reconstitution of triple helix takes place.

6 Concluding Remarks

In 1974 the late Paul Flory [388] categorised gels into four classes according to the mechanism of crosslinking. These were

1) Well-ordered lamellar structures, including gel mesophases
2) Covalent polymeric networks; completely disordered
3) Polymer networks formed through physical aggregation; predominantly disordered but with regions of local order
4) Particulate, disordered structures.

It should be clear from the preceding text that, within the constraints we ourselves defined in the Introduction, to limit our discussion to physical gels formed from biopolymer solutions, most of the systems discussed fall into one or other of the above divisions. Even covalent crosslinking can occur as in the case of the disulphide bonds formed at high temperatures in globular protein gels (Sect. 5.3.1). Further it could be argued that some of the systems fall into two or more such categories, agarose, for example, demonstrating features of both classes 1 and 3.

This perhaps illustrates the complication of work in this area, which combined with the background disciplines of the many investigators, (which naturally influence these workers toward particular techniques and consequent perceptions) has at times resulted in 'patchwork' approaches. Thus unfortunately the majority of studies have not been completely systematic. In view of the multifarious distance scales of interest, a multi-disciplinary, multi-technique approach is quite crucial if explanations and theories valid under a range of conditions are to be extracted. However, it should also be clear from the preceding text that there are honourable exceptions to this *caveat*, and the extensive studies on fibrin demonstrate this. In this context it is perhaps not coincidental that many of these have been carried out by J. D. Ferry, who following his very early review [3], might be termed the pioneer of such rigorous studies on biological gels.

At the same time prospects for further progress in the area are encouraging, and should be judged by the acceleration of interest by workers in synthetic polymer networks intrigued by (and maybe a little frustrated by) the complexities of behaviour exhibited. Again there is cause of cautious optimism here. Modern physics has given us a new constructive insight into the behaviour of systems by demonstrating that apparently distinct physical models, are, in the limit, governed by certain universalities of behaviour [389]. Provided we accept that within these there are classes of universality (*and* also clear anomalies of a chemical nature), this is a very useful philosophical approach, and one we ourselves have deliberately adopted by considering both components of the matrix, viz. techniques and systems. Herein techniques are meant to demonstrate the generalities and systems the specifics.

One topic which we have covered in considerable depth in this article, not only because of its great interest to us, but also because of its clear importance in the future development of rigorous physical theories, is the applicability of network theories to these systems. There are clearly two distinct viewpoints here, one belonging to those who accept that theories of rubber elasticity can never be applied to such systems, and one belonging to those who claim that, when modified, these approaches do afford us the opportunity to generate understanding. It should also be clear from the earlier text that, bearing in mind all the potential problems that could exist, we are firmly in the latter camp. However, even within this camp there are discrete factions, and a particularly controversial aspect is whether or not we should still assume the kT per network chain contribution of rubber elasticity. We have argued that, at substantial degrees of crosslinking, this cannot, in general, be true. However we cannot deny the truth of the complementary viewpoint, i.e. that just at, or very close to, the gel point, the length of these network chains must be extremely long, and thus, however persistent the chains, they should in this limit follow Gaussian statistics. What this implies is that, for example, the parameter 'a' in Eq. (25) should be itself a slowly increasing function of C/C_0.

Clearly the adoption of this more sophisticated model would require for its justification further experimental data, particularly from more systematic experiments, including those in which sharp discrete molecular weight fractions were employed, and as many techniques as possible brought to bear to characterise the density of crosslinks in an unequivocal way. Such a rigorous study is, unfortunately, extremely time consuming, but we hope that this Review will encourage more workers, to attack this very stimulating area, and hence bring to bear novel approaches to this topic.

Acknowledgements: It is a pleasure to acknowledge the many workers with whom we have corresponded during the production of this article, and who have generously sent us preprints of their work. Amongst those who have made particular contributions we mention Prof. Walther Burchard of Universitat Freiburg, Dr. Jean-Pierre Busnel, Universite du Maine, Profs. Manfred Gordon and Sir Sam Edwards of the University of Cambridge, and Prof. Ed Morris of Silsoe College, Cranfield Institute of Technology. We are grateful also to our colleagues within Unilever Research, notably Dr. Johannes Visser, Peter Wilding and also Dr. Alex Lips for his enthusiasm and continuing encouragement. We are, as ever, indebted to Lesley Linger and Helen Rose for their help in the preparation of the manuscript, and to our editor Prof. Karel Dušek at the I.M.C. Prague for his patience and support.

7 References

1. 'Polymer Networks', Adv. Polym. Sci. *44*, 1 (1982) (Dušek, K. Ed.)
2. Brit. Polym. J. *17* (2), 95–250 (1985) (Networks 84 Issue)
3. Ferry, J. D.: Adv. Protein Chem. *4*, 1 (1948)
4. Bisschops, J.: J. Polym. Sci. *17*, 89 (1955)
5. Agarwal, P. K., Lundberg, R. D.: Macromolecules *17*, 1928 (1984)
6. Atkins, E. D. T., Isaac, D. H., Keller, A.: J. Polym. Sci. (Polymer Physics Ed.) *18*, 71 (1980)
7. Atkins, E. D. T., Hill, M. J., Jarvis, D. A., Keller, A., Sarhene, E., Shapiro, J. S.: Colloid & Polymer Sci. *262*, 22 (1984)
8. Tsuchida, E., Abe, K.: Adv. Polym. Sci. *45*, 1 (1982)
9. Graessley, W. W.: Adv. Polym. Sci. *16*, 1 (1974)
10. Graessley, W. W.: Adv. Polym. Sci. *47*, 68 (1982)
11. Ferry, J. D.: Viscoelastic Properties of Polymers 3rd Edition, John Wiley, New York (N.Y.) 1980
12. Manson, J. A., Sperling, L. H.: Polymer Blends and Composites, Heyden, London 1976
13. Jennings, B. R. (Ed.): Electro-Optics and Dielectrics of Macromolecules and Colloids, Plenum Press, New York, 1979
14. Tsvetkov, V. N., Rjumtsev, E. I., Shtennikova, I. N.: in Liquid Crystalline Order in Polymers (Blumstein, A. Ed.), Academic Press, New York, 1978, p. 43
15. Gratzer, W. B.: in Poly-α-amino Acids (Fasman, G. D. Ed.), Marcel Dekker, New York, 1967, p. 177
16. Beychok, S.: in Poly-α-amino Acids (Fasman, G. D. Ed.), Marcel Dekker, New York, 1967, p. 293
17. Yang, J. T.: in Poly-α-amino Acids (Fasman, G. D. Ed.), Marcel Dekker, New York, 1967, p. 239
18. Schellman, J. A.: Acc. Chem. Res. *1*, 144 (1968)
19. Rees, D. A.: Pure Appl. Chem. *53*, 1 (1981)
20. Wilson Jr., E. B., Decius, J. C., Cross, P. C.: Molecular Vibrations, McGraw-Hill, New York, 1955
21. Miyazawa, T., Blout, E. R.: J. Amer. Chem. Soc. *83*, 712 (1961)
22. Nevskaya, N. A., Chirgadze, Y. N.: Biopolymers *15*, 637 (1976)
23. Chirgadze, Y. N., Nevskaya, N. A.: Biopolymers *15*, 607, 627 (1976)
24. Frushour, B. G., Koenig, J. L.: Biopolymers *13*, 1809 (1974)
25. Bryce, T. A., McKinnon, A. A., Morris, E. R., Rees, D. A., Thom, D. A.: Farad. Discuss. Chem. Soc. *57*, 221 (1974)
26. Ablett, S., Clark, A. H., Rees, D. A.: Macromolecules *15*, 597 (1982)
27. Rochas, C., Rinaudo, M., Vincendon, M.: Biopolymers *19*, 2165 (1980)
28. Rochas, C., Rinaudo, M., Vincendon, M.: Int. J. Biol. Macromol. *5*, 111 (1983)
29. Mackie, W., Perez, S., Rizzo, R., Taravel, F., Vignon, M.: Int. J. Biol. Macromol. *5*, 329 (1983)
30. Grasdalen, H., Smidsrød, O.: Macromolecules *14*, 229 (1981)
31. Belton, P. S., Morris, V. J., Tanner, S. F.: Int. J. Biol. Macromol. *7*, 53 (1985)
32. Ablett, S., Lillford, P. J., Baghdadi, S. M. A., Derbyshire, W.: J. Coll. Interface Sci. *67*, 355 (1978)
33. Maquet, J., Thevenau, H., Djabourov, M., Papon, P.: Int. J. Biol. Macromol. *6*, 162 (1984)
34. Burchard, W.: Brit. Polym. J. *17*, 154 (1985)
35. Clark, A. H., Lee-Tuffnell, C. D.: in Functional Properties of Food Macromolecules (Mitchell, J. R., Ledward, D. A. Eds.) Elsevier/Applied Science, Barking U.K., 1986, p. 203
36. Family, F., Landau, D. P. (Eds.): Kinetics of Aggregation and Gelation, Elsevier Science, Amsterdam 1984
37. Meakin, P.: Phys. Rev. A *27*, 1495 (1983)
38. Ball, R. C., Nauernberg, M., Witten, T. A.: Phys. Rev. A *29*, 2017 (1984)
39. Witten, T. A., Sander, L. M.: Phys. Rev. B *27*, 5686 (1983)
40. Burchard, W., Thurn, A., Wachenfeld, E.: in Physics of Finely Divided Matter (Boccara, N., Daoud, M. Eds.) Springer Proceedings in Physics Vol. V, Berlin 1985, p. 128; (and Burchard, W.: priv. commun.)

41. de Gennes, P.-G.: Scaling Concepts in Polymer Physics, Cornell University Press, Ithaca (N.Y.), 1979
42. Munch, J. P., Candau, S. J., Herz, J., Hild, G.: J. Phys. (Paris) *38*, 971 (1977)
43. Hecht, A. M., Geissler, E.: J. Phys. (Paris) *39*, 631 (1978)
44. Candau, S. J., Young, C. Y., Tanaka, T., Lemarechal, P., Bastide, J.: J. Chem. Phys. *70*, 4694 (1979)
45. Candau, S. J., Bastide, J., Delsanti, M.: Adv. Polym. Sci. *44*, 27 (1982)
46. Brown, W., Stilbs, P., Johnsen, R. M.: J. Polym. Sci. (Polym. Phys. Ed.) *20*, 1771 (1980)
47. Amis, E. J., Janmey, P. A., Ferry, J. D., Yu, H.: Macromolecules *16*, 441 (1983)
48. Burchard, W., Eisele, M.: Pure Appl. Chem. *56*, 1379 (1984)
49. ter Meer, H.-U., Burchard, W.: in Integration of Fundamental Polymer Science and Technology (Kleintjens, L. A., Lemstra, P. J. Eds.) Elsevier Applied Science, London 1986, p. 230
50. Schätzel, K.: in Optical Measurements in Fluid Mechanics 1985, (Richards, P. H. Ed.) Inst. Physics Conf. Ser. Vol. 77, Adam Hilger, Bristol, 1985, p. 175
51. Hervet, H., Leger, L., Rondelez, F.: Phys. Rev. Lett. *42*, 1681 (1979)
52. Chang, T., Yu, H.: Macromolecules *17*, 115 (1984)
53. Morris, V. J., Fancey, K. S.: Int. J. Biol. Macromol. *3*, 213 (1981)
54. Wun, K. L., Feke, G. T., Prins, W.: Farad. Discuss. Chem. Soc. *57*, 146 (1974)
55. Feke, G. T., Prins, W.: Macromolecules *7*, 527 (1974)
56. Key, P. Y., Sellen, D. B.: J. Polym. Sci. (Polymer Physics Ed.) *20*, 659 (1982)
57. Saio, K., Kamiya, M., Watanabe, T.: Agric. Biol. Chem. *33*, 1301 (1969)
58. Marshall, S. G., Vaisey, M.: J. Text. Stud. *3*, 173 (1972)
59. Szczesniak, A. S.: J. Text. Stud. *6*, 139 (1975)
60. Peniche-Covas, C. A. L., Dev, S. B., Gordon, M., Judd, M., Kajiwara, K.: Farad. Discuss. Chem. Soc. *57*, 165 (1974)
61. Richardson, R. K., Ross-Murphy, S. B.: Int. J. Biol. Macromol. *3*, 315 (1981)
62. Tokita, M., Niki, R., Hikichi, K.: J. Phys. Soc. Japan *53*, 480 (1984)
63. Gordon, M., Roberts, K. R.: Polymer *20*, 689 (1979)
64. Macosko, C. W.: Brit. Polymer J. *17*, 239 (1985)
65. Bibbo, M. A., Valles, E. M.: Macromolecules *17*, 360 (1984)
66. Richardson, R. K., Robinson, G., Ross-Murphy, S. B., Todd, S.: Polymer Bull. *4*, 541 (1981)
67. Djabourov, M., Maquet, J., Theveneau, H., Leblond, J., Papon, P.: Brit. Polymer J. *17*, 169 (1985)
68. Jen, C. J., McIntire, L. V., Bryan, J.: Arch. Biochim. Biophys. *216*, 126 (1982)
69. Jen, C. J., McIntire, L. V.: Biochim. Biophys. Acta *801*, 410 (1984)
70. Clark, A. H., Ross-Murphy, S. B.: Brit. Polymer J. *17*, 164 (1985)
71. Clark, A. H., Richardson, R. K., Robinson, G., Ross-Murphy, S. B., Weaver, A. C.: Prog. Food Nutrit. Sci. *6*, 149 (1982)
72. Adam, M., Delsanti, M., Durand, D.: Macromolecules *18*, 2285 (1985)
73. Whorlow, R. W.: Rheological Techniques, Ellis Horwood Chichester, U.K., 1980
74. Ellis, H. S., Ring, S. G.: Carbohdr.Polym. *5*, 201 (1985)
75. Mitchell, J. R.: in Proc. Int. Workshop on Plant Polysaccharides Structure and Function (Ed. Mercier, C., Rinaudo, M.) INRA/CNRS 1984, p. 93
76. Morris, E. R., Ross-Murphy, S. B.: Techniques in the Life Sciences, B *310*, 1 (1981)
77. Mooney, M.: J. Polym. Sci. *34*, 599 (1959)
78. Treloar, L. R. G.: The Physics of Rubber Elasticity, Clarendon Press, Oxford, 2nd Edition 1970
79. van den Tempel, M.: J. Colloid Sci. *16*, 284 (1961)
80. Mitchell, J. R., Blanshard, J. M. V.: J. Text. Studies *7*, 219; 341 (1976)
81. Braudo, E. E., Plashcina, I. G., Tolstoguzov, V. B.: Carbohydr. Polym. *4*, 23 (1984)
82. Mitchell, J. R.: J. Text. Studies *11*, 315 (1980)
83. Richardson, R. K., Ross-Murphy, S. B.: Brit. Polym. J. *13*, 11 (1981)
84. Eldridge, J. E., Ferry, J. D.: J. Phys. Chem. *58*, 992 (1954)
85. Bird, R. B., Armstrong, R. C., Hassager, O.: Dynamics of Polymeric Liquids, John Wiley, New York 1977
86. de Gennes, P.-G.: J. Chem. Phys. *55*, 572 (1971)
87. Doi, M., Edwards, S. F.: J. Chem. Soc. Faraday Trans. *2*, 74, 1789; 1802; 1818 (1978)

88. Klein, J., Fletcher, D., Fetters, L. J.: Faraday Symp. Chem. Soc. *18*, 159 (1983)
89. Baxandall, L. G.: Ph. D. Thesis, University of Cambridge, 1985
90. Edwards, S. F., Anderson, P. W.: J. Phys. F *6*, 1927 (1975)
91. Flory, P. J.: Trans. Farad. Soc. *56*, 722 (1960)
92. Tschoegl, N. W., Rinde, J. A., Smith, T. L.: Rheologica Acta *9*, 223 (1970)
93. Ross-Murphy, S. B., Todd, S.: Polymer *24*, 481 (1983)
94. Myers, F. S., Wenrick, J. D.: Rubber Chem. Technol. *47*, 1213 (1974)
95. McEvoy, H., Ross-Murphy, S. B., Clark, A. H.: Polymer *26*, 1483 (1985)
96. McEvoy, H., Ross-Murphy, S. B., Clark, A. H.: Polymer *26*, 1493 (1985)
97. Flory, P. J.: Proc. R. Soc. London A *351* (1666), *351* (1976)
98. Deam, R. T., Edwards, S. F.: Phil. Trans. A *280*, 317 (1976)
99. Mark, J. E.: Adv. Polym. Sci. *44*, 1 (1982)
100. Staverman, A. J.: Adv. Polym. Sci. *44*, 73 (1982)
101. Bailey, E., Mitchell, J. R., Blanshard, J. M. V.: Colloid & Polymer Sci. *255*, 856 (1977)
102. Mooney, M.: J. Appl. Phys. *11*, 582 (1940)
103. Blatz, P. I., Sharda, S. C., Tschoegl, N. W.: Trans. Soc. Rheol. *18*, 145 (1976)
104. Kilian, H.-G., Vilgis, Th.: Colloid & Polym. Sci. *262*, 15 (1984)
105. Doi, M., Kuzuu, N. Y.: J. Polym. Sci. (Polymer Physics Ed.) *18*, 409 (1980)
106. Williams, J. G.: Adv. Polym. Sci. *27*, 67 (1978)
107. Bucknall, C. B.: Adv. Polym. Sci. *27*, 121 (1978)
108. Smith, T. L. in: Rheology Vol V p. 127, Eirich, F. R. ed., Academic Press, New York, 1969
109. Flory, P. J.: Principles of Polymer Chemistry, Cornell University Press, Ithaca (N.Y.), 1953
110. Flory, P. J.: J. Amer. Chem. Soc. *63*, 3083, 3091 (1941)
111. Stockmayer, W. H.: J. Chem. Phys. *11*, 45 (1943)
112. Gordon, M., Ross-Murphy, S. B.: Pure Appl. Chem. *43*, 1 (1975)
113. Godovsky, Y. K.: Adv. Polym. Sci. *76*, 31 (1986)
114. Hermans, J. R.: J. Polym. Sci. Part A. *3*, 1859 (1965)
115. Gordon, M.: in Integration of Fundamental Polymer Science and Technology (Kleintjens, L. A., Lemstra, P. J. Eds.) Elsevier Applied Science, 1986, p. 167
116. Kilb, R. W.: J. Phys. Chem. *62*, 969 (1958)
117. Dušek, K., Gordon, M., Ross-Murphy, S. B.: Macromolecules *11*, 236 (1978)
118. Mark, J. E., Curro, J. G.: J. Chem. Phys. *79*, 5705 (1983)
119. Mark, J. E., Curro, J. G.: J. Chem. Phys. *81*, 6408 (1984)
120. Johnston, D. E.: J. Dairy. Res. *51*, 91 (1984)
121. Hirai, N.: Bull. Inst. Chem. Res. Kyoto *33*, 31 (1955)
122. Langley, N. R.: Macromolecules *1*, 348 (1968)
123. Oakenfull, D., Scott, A.: J. Food Sci. *49*, 1093 (1984)
124. Oakenfull, D.: J. Food Sci. *49*, 1103 (1984)
125. Stauffer, D., Coniglio, A., Adam, M.: Adv. Polym. Sci. *44*, 103 (1982)
126. Kajiwara, K., Burchard, W., Nerger, D., Dušek, K., Matejka, L., Tuzar, Z.: Makromol. Chem. *185*, 2453 (1984)
127. Mitchell, G. R., Windle, A. H.: Polymer *24*, 285 (1983)
128. Gronski, W., Stadler, R., Maldaner Jacobi, M.: Macromolecules *17*, 741 (1984)
129. Ferry, J. D., Eldridge, J. E.: J. Phys. Chem. *53*, 184 (1949)
130. Llorente, M. A., Mark, J. E.: J. Chem. Phys. *71*, 682 (1979)
131. Walter, A. T.: J. Polym. Sci. *13*, 207 (1954)
132. Watase, M., Nishinari, K.: Polym. Commun. *24*, 270 (1983)
133. Nishinari, K., Koide, S., Ogino, K.: J. Physique *46*, 793 (1985)
134. Hill, T. L.: J. Chem. Phys. *30*, 451 (1959)
135. Poland, D., Scheraga, H. A.: Theory of Helix-coil Transitions in Biopolymers, Academic Press, New York (NY) 1970
136. Dea, I. C. M., McKinnon, A. A., Rees, D. A.: J. Mol. Biol. *68*, 153 (1972)
137. Clark, A. H., Richardson, R. K., Ross-Murphy, S. B., Stubbs, J. M.: Macromolecules *16*, 1367 (1983)
138. Dušek, K.: J. Polym. Sci.: Part C *39*, 83 (1972)
139. Coniglio, A., Stanley, H. E., Klein, W.: Phys. Rev. B *25*, 6805 (1982)
140. Tanaka, T., Swislow, G., Ohmine, I.: Phys. Rev. Lett. *42*, 1556 (1979)

141. Flory, P. J.: J. Chem. Phys. *17*, 223 (1949)
142. Takahishi, A., Nakumura, T., Kagaura, I.: Polym. J. *3*, 207 (1972)
143. Donovan, J. W.: Biopolymers *18*, 263 (1979)
144. Blumstein, A. (Ed.): Liquid Crystalline Order in Polymers Academic Press, New York, 1978
145. Doi, M., Edwards, S. F.: J. Chem. Soc. Faraday Trans. *2*, 74, 560; 918 (1978)
146. Flory, P. J.: Proc. Roy. Soc. London A *234*, 73 (1956)
147. Warner, M., Flory, P. J.: J. Chem. Phys. *73*, 6327 (1980)
148. Khokhlov, A. R., Semenov, A. N.: Physica *112A*, 605 (1982)
149. Vorob'ev, V. I.: Biorheology *10*, 249 (1973)
150. Cahn, J. W.: J. Chem. Phys. *42*, 93 (1965)
151. Veis, A.: The Macromolecular Chemistry of Gelatin, Academic Press, London, 1964
152. Boedtker, H., Doty, P.: J. Amer. Chem. Soc. *78*, 4267 (1956)
153. Eagland, D., Pilling, G., Wheeler, R. G.: Farad. Discuss. Chem. Soc. *57*, 181 (1974)
154. Boedtker, H., Doty, P.: J. Phys. Chem. *58*, 968 (1954)
155. Smith, C. R.: J. Amer. Chem. Soc. *41*, 135 (1919)
156. Flory, P. J., Weaver, E. S.: J. Amer. Chem. Soc. *82*, 4518 (1960)
157. Harrington, W. F., von Hippel, P. H.: Arch. Biochem. Biophys. *92*, 100 (1961)
158. Avrami, M.: J. Chem. Phys. *7*, 1103 (1939); *8*, 212 (1940)
159. Harrington, W. F., Rao, N. V.: Biochemistry *9*, 3714 (1970)
160. Harrington, W. F., Karr, G. M.: Biochemistry *9*, 3725 (1970)
161. Yuan, L., Veis, A.: Biophys. Chem. *1*, 117 (1973)
162. Titova, E. F., Belavtseva, E. M., Braudo, E. E., Tolstoguzov, V. B.: Colloid & Polym. Sci. *252*, 497 (1974)
163. Tomka, I., Bohonek, J., Spuhler, A., Ribeaud, M.: J. Photograph. Sci. *23*, 97 (1975)
164. Finer, E. G., Franks, F., Phillips, M. C., Suggett, A.: Biopolymers *14*, 1995 (1975)
165. Djabourov, M., Papon, P.: Polymer *24*, 537 (1983)
166. Chatellier, J. Y., Durand, D., Emery, J. R.: Int. J. Biol. Macromol. *7*, 311 (1985)
167. Durand, D., Emery, J. R., Chatellier, J. Y.: Int. J. Biol. Macromol. *7*, 315 (1985)
168. Godard, P., Biebuyck, J. J., Daumerie, M., Naveau, H., Mercier, J. P.: J. Polym. Sci. Polym. Phys. Ed. *16*, 1817 (1978)
169. Jacobson, H., Stockmayer, W. H.: J. Chem. Phys. *18*, 1600 (1950)
170. te Nijenhuis, K.: Colloid & Polym. Sci. *259*, 522 (1981)
171. te Nijenhuis, K.: Colloid & Polym. Sci. *259*, 1017 (1981)
172. Rees, D. A.: Adv. Carbohydr. Chem. Biochem. *24*, 267 (1969)
173. Anderson, N. S., Campbell, J. W., Harding, M. M., Rees, D. A., Samuel, J. W. B.: J. Mol. Biol. *45*, 85 (1969)
174. Rees, D. A., Scott, W. E., Williamson, F. B.: Nature (London) *227*, 390 (1970)
175. Rees, D. A.: J. Chem. Soc. B., 877 (1970)
176. McKinnon, A. A., Rees, D. A., Williamson, F. B.: Chem. Comm., 701 (1969)
177. Jones, R. A., Staples, E. J., Penman, A.: Chem. Comm., 1608 (1973)
178. Reid, D. S., Bryce, T., Clark, A. H., Rees, D. A.: Farad. Discuss. Chem. Soc. *57*, 230 (1974)
179. Norton, I. T., Goodall, I. T., Morris, E. R., Rees, D. A.: Chem. Comm., 515 (1978)
180. Norton, I. T., Goodall, D. M., Morris, E. R., Rees, D. A.: J. Chem. Soc. Farad. Trans. 1 *79*, 2501 (1983)
181. Norton, I. T., Goodall, D. M., Morris, E. R., Rees, D. A.: J. Chem. Soc. Chem. Comm., 988 (1979)
182. Norton, I. T., Goodall, D. M., Morris, E. R., Rees, D. A.: J. Chem. Soc. Farad. Trans. 1 *79*, 2489 (1983)
183. Morris, E. R., Rees, D. A., Robinson, G.: J. Mol. Biol. *138*, 349 (1980)
184. Belton, P. S., Chilvers, G. R., Morris, V. J., Tanner, S. F.: Int. J. Biol. Macromol. *6*, 303 (1984)
185. Norton, I. T., Goodall, D. M., Morris, E. R., Rees, D. A.: J. Chem. Soc. Farad. Trans. 1 *79*, 2475 (1983)
186. Norton, I. T., Morris, E. R., Rees, D. A.: Carbohydr. Res. *134*, 89 (1984)
187. Grasdalen, H., Smidsrød, O.: Macromolecules *14*, 1842 (1981)
188. Smidsrød, O., Andresen, I.-L., Grasdalen, H., Larsen, B., Painter, T.: Carbohydr. Res. *80*, C11 (1980)
189. Smidsrød, O., Grasdalen, H.: Carbohydr. Polymers *2*, 270 (1982)

190. Morris, V. J., Belton, P. S.: J. Chem. Soc. Chem. Comm., 983 (1980)
191. Morris, V. J., Chilvers, G. R.: J. Sci. Food Agric. *32*, 1235 (1981)
192. Morris, V. J.: Int. J. Biol. Macromol. *4*, 155 (1982)
193. Morris, V. J., Chilvers, G. R.: Carbohdr. Polymers *3*, 129 (1983)
194. Hickson, T. L., Polson, A.: Biochem. Biophys. Acta *165*, 43 (1968)
195. Arnott, S., Fulmer, A., Scott, W. E., Dea, I. C. M., Moorhouse, R., Rees, D. A.: J. Mol. Biol. *90*, 269 (1974)
196. Liang, J. N., Stevens, E. S., Morris, E. R., Rees, D. A.: Biopolymers *18*, 327 (1979)
197. Obrink, B.: J. Chromatog. *37*, 329 (1968)
198. Pines, E., Prins, W.: Macromolecules *6*, 888 (1973)
199. Amsterdam, A., Er-El, Z., Shaltiel, S.: Arch. Biochem. Biophys. *171*, 673 (1975)
200. Laurent, T. C.: Biochim. Biophys. Acta *136*, 199 (1967)
201. Hayashi, A., Kinoshita, K., Kuwano, M., Nose, A.: Polymer J. *10*, 5 (1978)
202. Hayashi, A., Kinoshita, K., Yasueda, S.: Polym. J. *12*, 447 (1980)
203. Foord, S. A.: Ph. D. Thesis, University of Bristol, U.K., 1980
204. Atkins, E. D. T., Isaac, D. H., Keller, A., Miyasaka, K.: J. Polym. Sci. (Polym. Phys. Ed.) *15*, 211 (1977)
205. Smidsrød, O.: Farad. Discuss. Chem. Soc. *57*, 263 (1974)
206. Haug, A., Larsen, B., Smidsrød, O.: Acta Chem. Scand. *20*, 183 (1966)
207. Haug, A., Larsen, B., Smidsrød, O.: Acta Chem. Scand. *23*, 691 (1967)
208. Smidsrød, O., Haug, A.: Acta Chem. Scand. *22*, 797 (1968)
209. Smidsrød, O.: Carbohydr. Res. *13*, 359 (1970)
210. Mackie, W., Noy, R., Sellen, D. B.: Biopolymers *19*, 1839 (1980)
211. Whittington, S. G.: Biopolymers *10*, 1617 (1971)
212. Kohn, R., Larsen, B.: Acta Chem. Scand. *26*, 2455 (1972)
213. Kohn, R.: Pure Appl. Chem. *42*, 371 (1975)
214. Atkins, E. D. T., Nieduszynski, I. A., Mackie, W., Parker, K. D., Smolko, E. E.: Biopolymers *12*, 1865; 1879 (1973)
215. Mackie, W.: Biochem. J. *125*, 89 (1971)
216. Morris, E. R., Rees, D. A., Thom, D.: J. Chem. Soc. Chem. Comm., 245 (1973)
217. Grant, G. T., Morris, E. R., Rees, D. A., Smith, P. J. C., Thom, D.: FEBS Letts. *32*, 195 (1973)
218. Morris, E. R., Rees, D. A., Thom, D., Boyd, J.: Carbohydr. Res. *66*, 145 (1978)
219. Smidsrød, O., Haug, A.: Acta Chem. Scand. *26*, 79 (1972)
220. Segeren, A. J. M., Boskamp, J. V., van den Tempel, M.: Farad. Discuss. Chem. Soc. *57*, 255 (1974)
221. Nevell, T. P., Zeronian, S. H. (Eds.): Cellulose Chemistry and its Applications, Ellis Horwood, Chichester U.K., 1985
222. Klug, E. D.: J. Polym. Sci. C *36*, 491 1971
223. Sarkar, E. D.: J. Appl. Polym. Sci. *24*, 1073, 1979
224. Werbowyj, R. S., Gray, D. G.: Mol. Cryst. Liq. Cryst. *34*, 97, 1976
225. Laivins, G. V., Gray, D. G.: Macromolecules *18*, 1753, 1985
226. Conio, G., Bianchi, E., Ciferri, A., Tealdi, A., Aden, M. A.: Macromolecules *16*, 1264 (1983)
227. Morris, E. R., Gidley, M. J., Murray, E. J., Powell, D. A., Rees, D. A.: Int. J. Biol. Macromol. *2*, 327 (1980)
228. Davis, M. A. F., Gidley, M. J., Morris, E. R., Powell, D. A., Rees, D. A.: Int. J. Biol. Macromol. *2*, 330 (1980)
229. Gidley, M. J., Morris, E. R., Murray, E. J., Powell, D. A., Rees, D. A.: J. Chem. Soc. Chem. Comm., 990 (1979)
230. Walkinshaw, M. D., Arnott, S.: J. Mol. Biol. *153*, 1055 (1981)
231. Walkinshaw, M. D., Arnott, S.: J. Mol. Biol. *153*, 1075 (1981)
232. Morris, E. R., Powell, D. A., Gidley, M. J., Rees, D. A.: J. Mol. Biol. *155*, 507 (1982)
233. Powell, D. A., Morris, E. R., Gidley, M. J., Rees, D. A.: J. Mol. Biol. *155*, 517 (1982)
234. Dea, I. C. M., Morrison, A.: Adv. Carbohydr. Chem. Biochem. *31*, 241 (1975)
235. Courtois, J. E., LeDizet, P.: Bull. Soc. Chim. Biol. *52*, 15 (1970)
236. McCleary, B. V., Clark, A. H., Dea, I. C. M., Rees, D. A.: Carbohydr. Res. *139*, 237 (1985)

237. Morris, E. R., Cutler, A. N., Ross-Murphy, S. B., Rees, D. A., Price, J.: Carbohdr. Polym. 1, 5 (1981)
238. Robinson, G., Ross-Murphy, S. B., Morris, E. R.: Carbohydr. Res. *107*, 17 (1982)
239. Banks, W., Greenwood, C. T.: Starch and its Components, Edinburgh University Press, Edinburgh, 1975
240. Ring, S. G., Stainsby, G.: Proc. Food Nutrit. Sci. *6*, 323 (1982)
241. Miles, M. J., Morris, V. J., Ring, S. G.: Carbohydr. Polym. *4*, 73 (1984)
242. Morris, E. R.: in NATO Adv. Study Inst. Ser. Ser. C. 1978 C50 (Tech. Appl. Fast React. Solution) 1979, p. 379
243. Ring, S. G.: Int. J. Biol. Macromol. *7*, 253 (1985)
244. Sandford, P. A.: Adv. Carbohydr. Chem. Biochem. *36*, 265 (1979)
245. Morris, E. R., Rees, D. A., Young, G., Darke, A.: J. Mol. Biol. *110*, 1 (1977)
246. Norton, I. T., Goodall, D. M., Frangou, S. A., Morris, E. R., Rees, D. A.: J. Mol. Biol. *175*, 371 (1984)
247. Nisbet, B. A., Sutherland, I. W., Bradshaw, I. J., Kerr, M., Morris, E. R., Shepperson, W. A.: Carbohydr. Polym. *4*, 377 (1984)
248. Frangou, S. A., Morris, E. R., Rees, D. A., Richardson, R. K., Ross-Murphy, S. B.: J. Polym. Sci. (Polym. Lett. Ed.) *20*, 531 (1982)
249. Morris, V. J., Franklin, D., I'Anson, K.: Carbohydr. Res. 121, 13 (1983)
250. Brownsey, G. J., Chilvers, G. R., I'Anson, K., Morris, V. J.: Int. J. Biol. Macromol. *6*, 211 (1984)
251. Hascall, V. C.: in Biology of Carbohydrates Vol. 1 (Ginsburg, V., Robbins, P. Eds.), John Wiley, New York, 1981, p. 1
252. Morris, E. R., Rees, D. A., Welsh, E. J.: J. Mol. Biol. *138*, 383 (1980)
253. Welsh, E. J., Rees, D. A., Morris, E. R., Madden, J. K.: J. Mol. Biol. *138*, 375 (1980)
254. Cleland, R. L.: Arch. Biochem. Biophys. *180*, 57 (1977)
255. Allen, A.: Trends. Biochem. Sci. *8*, 169, 1983
256. Elstein, M., Parke, D. V.: Mucus in Health and Disease Plenum Press, New York, 1977; Chantler, E., Elder, J., Elstein, M.: Mucus in Health and Disease II, Plenum Press, New York, 1982
257. Snary, D., Allen, A., Pain, R. H.: Biochem. Biophys. Res. Commun. *40*, 844 (1970)
258. Scawen, M., Allen, A.: Biochem. J. *163*, 363 (1977)
259. Allen, A., Pain, R. H., Snary, D.: Farad. Discuss. Chem. Soc. *57*, 210 (1974)
260. Denton, R., Forsman, W., Hwang, S. H., Litt, M., Miller, C. E.: Amer. Rev. Respiratory Diseases *98*, 380 (1968)
261. Litt, M., Khan, M. A., Wolf, D. P.: Biorheology *13*, 37 (1976)
262. Martin, G. P., Marriott, C., Kellaway, I. W.: Gut *19*, 103 (1978)
263. Bell, A. E., Allen, A., Morris, E. R., Ross-Murphy, S. B.: Int. J. Biol. Macromol. *6*, 309 (1984)
264. Bell, A. E., Sellers, L. A., Allen, A., Cunliffe, W. J., Morris, E. R., Ross-Murphy, S. B.: Gastroenterology *88*, 269 (1985)
265. Sellers, L. A.: Ph.D. Thesis, University of Newcastle-upon-Tyne, 1985 and to be published
266. Newlin, T. E., Lovell, S. E., Saunders, P. R., Ferry, J. D.: J. Colloid Sci. *17*, 10 (1962)
267. van den Tempel, M.: in Rheometry-Industrial Applications (Walters, K. Ed.), John Wiley, New York, 1980, p. 179
268. Ross-Murphy, S. B.: in Biophysical Methods in Food Research — Critical Reports on Applied Chemistry Vol. 5 (Chan, H. W. -S. Ed.) Blackwell, Oxford, U.K., 1984, p. 138
269. Whitcomb, P. J., Macosko, C. W.: J. Rheol. *22*, 493 (1978)
270. Wissbrun, K. F.: J. Rheol. *25*, 619 (1981)
271. Jansson, P. E., Kenne, L., Lindberg, B.: Carbohydr. Res. *45*, 275 (1975)
272. Milas, M., Rinaudo, M.: Carbohydr. Res. *76*, 189 (1979)
273. Moorhouse, R., Walkinshaw, M. D., Arnott, S.: in Extracellular Microbial Polysaccharides (Sandford, P. A., Laskin, A. Ed.) ACS Symp. Ser. *45*, 90 (1977)
274. Morris, E. R.: in Extracellular Microbial Polysaccharides (Sandford, P. A., Laskin, A. Ed.) ACS Symp. Ser. *45*, 81 (1977)
275. Sato, T., Norisuye, T., Fujita, H.: Polymer J. *16*, 341 (1984)
276. Sato, T., Kojima, S., Norisuye, T., Fujita, H.: Polymer J. *16*, 423 (1984)
277. Paradossi, G., Brant, D. A.: Macromolecules *15*, 874 (1982)

278. Coviello, T., Kajiwara, K., Burchard, W., Dentini, M., Crescenzi, V.: Macromolecules *19*, 2826 (1986)
279. Muller, G., Anrhourrache, M., Lecourtier, J., Chauveteau, G.: Int. J. Biol. Macromol. *8*, 167 (1986)
280. Ross-Murphy, S. B., Morris, V. J., Morris, E. R.: Farad. Symp. Chem. Soc. *18*, 115 (1983)
281. Jamieson, A. M., Southwick, J. G., Blackwell, J.: Farad. Symp. Chem. Soc. *18*, 131 (1983)
282. Maret, G., Milas, M., Rinaudo, M.: Polym. Bull. *4*, 291 (1981)
283. Gibbs, D. A., Merrill, E. W., Smith, K. A., Balazs, E. A.: Biopolymers *6*, 777 (1968)
284. Lowey, S.: in Polymerization in Biological Systems CIBA Foundation Symposium, Elsevier-Excerpta Medica North Holland 1972, p. 217
285. Sattilaro, R. F., Dentler, W. L., Le Cluyse, E. L.: J. Cell Biology *90*, 467 (1981)
286. Hanson, J., Lowy, J.: J. Mol. Biol. *6*, 46 (1963)
287. Kasai, M., Kawashima, H., Oosawa, F.: J. Polym. Sci. *44*, 51 (1960)
288. Maruyama, K., Kaibara, M., Fukada, E.: Biochim. Biophys. Acta *371*, 20 (1974)
289. Zaner, K. S., Stossel, T. P.: J. Biol. Chem. *258*, 11004 (1983)
290. Jain, S., Cohen, C.: Macromolecules *14*, 759 (1981)
291. Carlson, F. D., Fraser, A. B.: J. Mol. Biol. *89*, 273 (1974)
292. Fujime, S., Ishiwata, S.: J. Mol. Biol. *62*, 251 (1971)
293. Ishiwata, S., Fujime, S.: J. Mol. Biol. *68*, 511 (1972)
294. Johnson, K. A., Borisy, G. G.: J. Mol. Biol. *117*, 1 (1977)
295. Murphy, D. B., Johnson, K. A., Borisby, G. G.: J. Mol. Biol. *117*, 33 (1977)
296. Mandelkow, E. M., Harmsen, A., Mandelkow, E., Bordas, J.: Nature (London) *287*, 595 (1980)
297. Nelson, R. L., McIntire, L. V., Brinkley, W. R.: J. Cell. Biol. in press
298. Finch, J. T., Perutz, M. F., Bertles, J. F., Dobler, J.: Proc. Nat. Acad. Sci. USA *70*, 718 (1973)
299. Rosen, L. S., Magdoff-Fairchild, B.: J. Mol. Biol. *183*, 565 (1985)
300. Josephs, R., Jarosch, H. S., Edelstein, S. J.: J. Mol. Biol. *102*, 409 (1976)
301. Dykes, G. W., Crepeau, R. H., Edelstein, S. J.: J. Mol. Biol. *130*, 451 (1979)
302. Miller, W. G., Chakrabarti, S., Seibel, K. M.: in Microdomains in Polymer Solutions (Dubin, P. L. Ed.) Plenum Press, New York, 1985, p. 143
303. Miller, W. G., Kou, L., Tokyama, K., Voltaggio, V.: J. Polym. Sci. (Polym. Sympos.) *65*, 91 (1978)
304. Koltun, W. L., Waugh, D. F., Bear, R. S.: J. Amer. Chem. Soc. *76*, 413 (1954)
305. Burke, M. J., Rougvie, M. A.: Biochemistry *11*, 2435 (1972)
306. Clark, A. H., Judge, F. J., Richards, J. B., Stubbs, J. M., Suggett, A.: Int. J. Peptide Protein Res. *17*, 380 (1981)
307. Clark, A. H., Saunderson, D. H. P., Suggett, A.: Int. J. Peptide Protein Res. *17*, 353 (1981)
308. Gratzer, W. B., Beaven, G. H., Rattle, H. W. E., Bradbury, E. M.: Eur. J. Biochem. *3*, 276 (1968)
309. Beaven, G. H., Gratzer, W. B., Davies, H. G.: Eur. J. Biochem. *11*, 37 (1969)
310. Scheraga, H. A., Laskowski, M.: Adv. Protein. Chem. *12*, 1 (1957)
311. Doolittle, R. F.: Adv. Protein Chem. *27*, 1 (1973)
312. Müller, M., Burchard, W.: in Fibrinogen — Recent Biochemical and Medical Aspects (Henschen, A., Graeff, H., Lottspeich, F. Eds.) W. de Gruyter, Berlin, 1982, p. 29
313. Koppel, G.: Nature (London) *212*, 1608 (1966)
314. Marguerie, G., Pouit, L., Suscillon, M.: Thromb. Res. *3*, 675 (1973)
315. Hall, C. E., Slayter, H. S.: J. Biophys. Biochem. Cytol. *5*, 11 (1959)
316. Stryer, L., Cohen, C., Langridge, R.: Nature (London) *197*, 793 (1963)
317. Roska, F. J., Ferry, J. D., Lin, J. S., Anderegg, J. W.: Biopolymers *21*, 1833 (1982)
318. Bang, N. U.: Thromb. Diath. Haemorrh. Suppl. *13*, 73 (1964)
319. Kay, D., Cuddington, B. J.: Brit. J. Haematol. *13*, 341 (1967)
320. Lederer, K., Hammel, R.: Makromol. Chem. *176*, 2619 (1975)
321. Müller, M., Burchard, W.: Biochim. Biophys. Acta *537*, 208 (1978); Int. J. Biol. Macromol. *3*, 71 (1981)
322. R. C. Williams, Müller, M., Burchard, W.: J. Mol. Biol. *150*, 399 (1981)
323. Bachmann, L., Schmitt-Fumian, W. W., Hammer, R., Lederer, K.: Makromol. Chem. *176*, 2603 (1975)
324. Bailey, K., Astbury, W. J., Rudall, K. M.: Nature (London) *151*, 716 (1943)

325. Burchard, W., Müller, M.: Int. J. Biol. Macromol. *2*, 225 (1980)
326. Shah, G. A., Ferguson, I. A., Dhall, T. Z., Dhall, D. P.: Biopolymers *21*, 1037 (1982)
327. Müller, M. F., Ris, H., Ferry, J. D.: J. Mol. Biol. *174*, 369 (1984)
328. Ferry, J. D., Morrison, P. R.: J. Amer. Chem. Soc. *69*, 388 (1947)
329. Ferry, J. D., Miller, M., Shulman, S.: Arch. Biochem. Biophys. *34*, 424 (1951)
330. Roberts, W. W., Kramer, O., Rosser, R. W., Nestler, F. H. M., Ferry, J. D.: Biophys. Chem. *1*, 152 (1974)
331. Gerth, C., Roberts, W. W., Ferry, J. D.: Biophys. Chem. *2*, 208 (1974)
332. Nelb, G. W., Gerth, C., Ferry, J. D.: Biophys. Chem. *5*, 377 (1976)
333. Nelb, G. W., Kamykowski, G. W., Ferry, J. D.: Biophys. Chem. *13*, 15 (1981)
334. Janmey, P. A., Amis, E. J., Ferry, J. D.: J. Rheol. *27*, 135 (1983)
335. Kinsella, J. E.: CRC Crit. Rev. Food Sci. Nutr., 1976, p. 219
336. Astbury, W. J., Dickinson, S., Bailey, K.: Biochem. J. *29*, 2351 (1935)
337. Barbu, E., Joly, M.: Farad. Discuss. Chem. Soc. *13*, 77 (1953)
338. Kratochvil, P., Munk, P., Bartl, P.: Coll. Czech. Chem. Commun. *26*, 945 (1961)
339. Kratochvíl, P., Munk, P., Sedlacék, B.: Coll. Czech. Chem. Commun. *26*, 2806 (1961)
340. Kratochvíl, P., Munk, P., Sedlacék, B.: Coll. Czech. Chem. Commun. *27*, 115 (1962)
341. Kratochvíl, P., Munk, P., Sedlacék, B.: Coll. Czech. Chem. Commun. *27*, 788 (1962)
342. Tombs, M. P.: Proteins as Human Foods (Lawrie, R. A. Ed.) Butterworths, London 1970, p. 126
343. Clark, A. H., Tuffnell, C. D.: Int. J. Peptide Protein Res. *16*, 339 (1980)
344. Hermansson, A. M., Buchheim, W.: J. Colloid Interface Sci. *81*, 519 (1981)
345. Bikbov, T. M., Grinberg, V. Ya., Antonov, Yu. A., Tolstoguzov, V. B., Schmandke, H.: Polym. Bull. *1*, 865 (1979)
346. Bikbov, T. M., Grinberg, V. Ya., Schmandke, H., Chaika, T. S., Vaintraub, I. A., Tolstoguzov, V. B.: Colloid & Polymer Sci. *259*, 536 (1981)
347. Hermansson, A.-M.: J. Food. Sci. *47*, 1960 (1982)
348. Joly, M.: A Physico-Chemical Approach to the Denaturation of Proteins, Academic Press, London 1965
349. Timasheff, S. N., Susi, H., Stevens, L.: J. Biol. Chem. *242*, 5467 (1967)
350. van Kleef, F. S. M., Boskamp, J. V., van den Tempel, M.: Biopolymers *17*, 225 (1978)
351. Payens, T. A. J.: J. Dairy Res. *46*, 291 (1979)
352. Walstra, P.: J. Dairy Res. *46*, 317 (1979)
353. Stothart, P. H., Cebula, D. J.: J. Mol. Biol. *160*, 391 (1982)
354. Heertje, I., Visser, J., Smits, P.: Food Microstructure *4*, 267 (1985)
355. Payens, T. A. J., Wiersma, A. K., Brinkhuis, J.: Biophys. Chem. *6*, 253 (1977)
356. Payens, T. A. J.: Biophys. Chem. *6*, 263 (1977)
357. Payens, T. A. J.: Farad. Discuss. Chem. Soc. *65*, 164 (1978)
358. Payens, T. A. J., Wiersma, A. K.: Biophys. Chem. *11*, 137 (1980)
359. Roefs, S. P. F. M.: Structure of Acid Casein Gels, Thesis, Agricultural University, Wageningen, Netherlands, 1986
360. von Smoluchowski, M.: Z. Physik. Chem. *92*, 129 (1917)
361. Dalgleish, D. G.: J. Dairy Res. *45*, 653 (1979)
362. Dalgleish, D. G.: Biophys. Chem. *11*, 147 (1980)
363. Dalgleish, D. G.: J. Dairy Res. *47*, 231 (1980)
364. Darling, D. F., van Hooydonk, A. C. M.: J. Dairy. Res. *48*, 189 (1981)
365. Dalgleish, D. G.: J. Dairy. Res. *50*, 331 (1983)
366. Culioli, J., Sherman, P.: J. Text. Studies *9*, 257 (1978)
367. Johnston, D. E.: Milchwissenschaft *39*, 405 (1984)
368. Payens, T. A. J.: Neth. Milk Dairy J. *38*, 195 (1984)
369. van Vliet, T., Walstra, P.: Neth. Milk Dairy J. *39*, 115 (1985)
370. Parker, T. G., Dalgleish, D. G.: J. Dairy Res. *44*, 79; 85 (1977)
371. Parker, T. G., Dalgleish, D. G.: Biopolymers *16*, 2533 (1977)
372. Dalgleish, D. G., Patterson, E., Horne, D. S.: Biophys. Chem. *13*, 307 (1981)
373. Wee, E. L., Miller, W. G.: J. Phys. Chem. *75*, 1446 (1971)
374. Miller, W. G., Wu, L. L., Wee, E. L., Santee, G. L., Rai, J. H., Goebel, K. D.: Pure Appl. Chem. *38*, 37 (1974)

375. Tohyama, K., Miller, W. G.: Nature (London) *289*, 813 (1981)
376. Cahn, J. W., Hilliard, J. E.: J. Chem. Phys. *31*, 688 (1959)
377. Samejima, K., Hashimoto, K., Yasui, T., Fukuzawa, T.: J. Food. Sci. *34*, 242 (1969)
378. Samejima, K.: J. Coll. Dairying *7*, 143 (1978)
379. Ishioroshi, M., Samejima, K., Yasui, T.: J. Food. Sci. *44*, 1280 (1979)
380. Yasui, T., Ishioroshi, M., Nakano, H., Samejima, K.: J. Food Sci. *44*, 1201 (1979)
331. Ishioroshi, M.. Samejima, K., Yasui, T.: Agric. Biol. Chem. *47*, 2809 (1983)
382. Wright, D. J., Wilding, P.: J. Sci. Food Agric. *35*, 357 (1984)
383. Samejima, K., Ishioroshi, M., Yasui, T.: Agric. Biol. Chem. *47*, 2373 (1983)
384. Hermansson, A.-M., Harbitz, O., Langton, M.: J. Sci. Food. Agric. *37*, 69 (1986)
385. Samejima, K., Ishioroshi, M., Yasui, T.: J. Food. Sci. *46*, 1412 (1981)
386. Egelansdal, B., Fretheim, K., Harbitz, O.: J. Sci. Food Agric. *37*, 944 (1986)
387. Egelansdal, B., Fretheim, K., Samejima, K.: J. Sci. Food Agric. *37*, 915 (1986)
388. Flory, P. J.: Farad. Discuss. Chem. Soc. *57*, 7 (1974)
389. Stauffer, D.: Phys. Reports *54*, 1 (1979)

Editor: K. Dušek
Received August 6, 1986

Author Index Volumes 1–83

Subject Index

GPSR Compliance

The European Union's (EU) General Product Safety Regulation (GPSR) is a set of rules that requires consumer products to be safe and our obligations to ensure this.

If you have any concerns about our products, you can contact us on ProductSafety@springernature.com

In case Publisher is established outside the EU, the EU authorized representative is:

Springer Nature Customer Service Center GmbH
Europaplatz 3
69115 Heidelberg, Germany

Batch number: 10620881

Printed by Printforce, the Netherlands